JUSTICE DENIED

JUSTICE DENIED

The Role of Forensic Science in the Miscarriage of Justice

David Klatzow

Published by Zebra Press
an imprint of Random House Struik (Pty) Ltd
Company Reg. No. 1966/003153/07
Estuaries No. 4, Oxbow Crescent, Century Avenue,
Century City, Cape Town, 7441
PO Box 1144, Cape Town, 8000, South Africa

www.zebrapress.co.za

First published 2014
Reprinted in 2014

3 5 7 9 10 8 6 4 2

PUBLISHER: Marlene Fryer
MANAGING EDITOR: Ronel Richter-Herbert
EDITOR: Beth Lindop
PROOFREADER: Bronwen Leak
COVER DESIGN: Michiel Botha
TEXT DESIGN: Jacques Kaiser
TYPESETTER: Martin Endemann/
Monique van den Berg

Set in 11.5 pt on 15.5 pt Granjon

Printed and bound by Interpak Books, Pietermaritzburg, South Africa

ISBN 978 1 77022 694 4 (print)
ISBN 978 1 77022 695 1 (ePub)
ISBN 978 1 77022 696 8 (PDF)

CONTENTS

Acknowledgements vii
Introduction 1

1. Sir Bernard Spilsbury: 'That Incomparable Witness' 7
2. 'Don't speak to that bastard; the state is gunning for him' 22
3. How Cool Is Cool? 30
4. The Fickle Fingers of Fate 46
5. The Case of Shirley McKie 53
6. The Mayfield Debacle and Other Disasters 71
7. The Use and Abuse of Statistics in the Courtroom 83
8. The Whole Truth and Nothing but the Truth 95
9. Beyond Reasonable Doubt 99
10. The *CSI* Effect and Reality 107
11. Bombing Out in Birmingham 115
12. The Burning Question 121
13. One for the Road 131
14. The War on Drugs 145
15. Biting the Bullet 154
16. Guns for Hire? 168
17. *Fiat justitia ruat caelum*: Let justice be done though the heavens fall 181
18. 'But Doctor, why would this policeman lie?' 195
19. Bad Apples, Bad Orchards or Bad Soil – Or All Three? 204
20. A Tale of Two Cultures 218

Epilogue 229
Appendix A 231
Appendix B 233
Appendix C 236
Appendix D 240

Appendix E 244
References 251
Index 265

ACKNOWLEDGEMENTS

The modest success of my previous book about my thirty years in forensic science emboldened me to write in this book about some of the problems I have encountered along the way. Like all narratives, the formative influences in my life paved the way for my subsequent career in forensic science.

One can never overstate the value of empathetic teachers who awakened dormant interests in the budding student. One of the first was a teacher at St Martin's School in Rosettenville. His name was Edwin Ainscow and he beat the notion of the balanced chemical equation into my head. Subsequently, a teacher called Rob Taylor aroused my interest in organic chemistry, and biochemistry in particular. In *Steeped in Blood* I mentioned my English teachers at Nigel English Medium High School: Jean Cameron and Dr Gevers. English teachers have the most amazing opportunity to mould the young mind in so many ways, and these two were exceptional in the breadth and depth of their understanding, and in their willingness to impart such learning.

Having left high school, I came under the influence of many men at university, some good and some indifferent. I mentioned in the introduction to *Steeped in Blood* my immense debt to Professor Jack Allan, who guided me in anatomical pathways. An oversight in the previous book was my omission of Professor Charles Isaacson, who took it upon himself to teach me the rudiments of pathology. They were enjoyable lessons of enormous value to me in the years to come and he, I am sure, is unaware of how much of a debt of gratitude I owe him. He has remained a steadfast friend, humble, erudite and caring.

My friend Solly (Stephan) van Nieuwenhuizen, now a silk at the Johannesburg Bar, has been a constant friend in need. He has, over the nearly thirty years of our friendship, been a consistent if sometimes irascible teacher of the law to me. His friendship has been invaluable.

My interest in serology was encouraged in the early stages of my PhD by husband-and-wife team Gerard and Dell Vos. He was a serologist of international repute who worked for, and was unappreciated by, the Natal Blood Transfusion Service, run at the time by Peter Brain. Gerard was quick to appreciate the role of the cell surface in cancer research, something that made of him a rather rare bird in South Africa in the 1970s. His encouragement and help was invaluable in my early career, especially his support of an intellectual nature.

During my time as a postgraduate student in the Department of Surgery, I was encouraged and supported by a number of senior members of staff. Professor Don Heynes, who was head of the ear nose and throat division, was always available and encouraging. He included me each week in his ward rounds and gave me valuable insights into the clinical aspects of the cancer research that I was engaged in. The same was true of Professor Noll van Blerk, head of the urology department. Professor D.J. du Plessis, who was head of surgery, was a strict but fair disciplinarian who would open doors to assist students he thought worth helping. God help you if you crossed him; his wrath was all-powerful and far reaching. I found him helpful, supportive and encouraging. He called me into his office at 6:30 one morning and gave me a lecture on how privileged I was to be given the opportunity to lay one brick in the edifice of knowledge. He ended by saying to me, 'Don't try to build the entire building.'

This section would be incomplete without mentioning the able assistance that I have received from my assistant, Natalie Salter-Buffard. At all times, her high standards and professionalism has made every bit of difference. Her sometimes Cerberus-like treatment of some of the strange clients who call on me has been both helpful and entertaining. Her untiring help with the preparation of this book has been invaluable. Thanks, Natalie.

To the staff of Zebra Press, especially Marlene Fryer, who has kept up the pressure, and Beth Lindop, who has tied up all the loose ends, thanks for your patience and persistence.

INTRODUCTION

The concept of justice goes far back into the dawn of recorded time. The ancient Egyptian tombs show ample evidence of weighing the soul of the dead, an idea illustrated in the *Book of the Dead*, an ancient Egyptian funerary text designed to assist souls through the underworld to the afterlife. In modern times, the concept of justice is represented by the Roman goddess Justitia, or Lady Justice. She is usually portrayed as a blindfolded woman with a sword in one hand and scales in the other. The symbolism is unequivocal and simple: justice should be blind; your class, colour, wealth and rank should play no part. The scales tell us that evidence must be weighed, the issues viewed in a balanced way. (Lawyers speak of something being proved 'on the balance of probabilities'.) The whiteness and simplicity of her toga-like gown speaks to the need for justice to encompass not only a civilised legal component but also a greater moral purpose.

Our judges are steeped in this tradition. They are referred to as 'the *honourable* Mr Justice So-and-so'. They wear robes that set them apart from the masses, and in the High Courts are addressed as 'My Lord' or 'My Lady'. The courts resound with obsequious phrases, such as 'If your Lordship pleases' and 'May it please the Court, your Ladyship'. Invariably the judicial bench is physically elevated well above the rest of the court.

The purpose of a court is to see to it that some form of justice is meted out to those unfortunate enough to end up in front of the judge presiding. Increasingly it is accepted that the law should be tempered with equity to better ensure that justice is served.

Does this happen in practice? I would venture to say not often. The legal process is a very expensive one; often justice is not achieved because the parties are not able to afford the costs. Costs of advocates and their instructing attorneys mount up alarmingly. If a senior counsel, together with junior counsel, is employed, the costs rapidly become stratospheric. This situation is made worse by the fact that the loser in most court actions is burdened still further with a costs order for the winning side's expenses in addition to his own. Should a party be acquitted in a criminal case, no costs order is awarded against the state, as would occur in a civil matter. As several of the cases discussed in this book reveal, the accused also often lacks the necessary finance to employ adequate technical forensic expertise to challenge the vast resources of the state.

In his book *The Log from the Sea of Cortez*, John Steinbeck paints a poignant picture of his friend Ed Ricketts, the famous doctor from Steinbeck's *Cannery Row* and *Sweet Thursday*. Ricketts went to the Superior Court in Salinas, California, seeking recompense for a fault of the electricity company whose negligence had caused his laboratory to burn down. In the ensuing court case, the doctor examined the legal system in the same way as he would have examined a new species of sea life as a biologist, and came to the conclusion that neither side wanted the truth. In fact, they had an aversion to the truth; they simply wanted to win.

Nothing has changed since that book was published in 1951 and, indeed, court cases both before and since then have all too often been mere parodies of justice. Frequently the courts are a forum where the poor seek justice, often in vain, and the great and the good make fools of themselves – as this book will show. It is not just in court that this happens: history has brought to light the foolishness of certain men who were renowned as experts in their subjects. In 1781, for example,

Sir William Herschel, a noted astronomer of the English royal court, was of the opinion that 'the sun is richly populated with inhabitants'. In a similarly short-sighted comment, Lord Kelvin, who was as far up the scientific ladder as it was possible to get, said in the early twentieth century: 'There is nothing new to be discovered in physics now. All that remains is more and more precise measurement.' That, of course, was before Einstein, Max Planck and the dawn of the atomic age of physics, which turned physics on its head. New Zealand–born physicist Ernest Rutherford, at the time head of the Cavendish laboratory in Cambridge and Nobel prize winner in Physics, declared: 'The energy produced by the breaking down of the atom is a very poor kind of thing. Anyone who expects the source of power from transformation of these atoms is talking moonshine.'

These foolish statements by eminent men are a source of gentle entertainment to us a hundred years later, but when foolish statements are uttered in a court of law by someone held in high esteem, for whatever reason, by the court, great harm and great injustice can be done. These kinds of declarations should serve as a lesson that renowned men can be way off the mark, and that eminence is not a sufficient basis for the acceptance of an opinion.

James E. Starrs of George Washington University has even gone so far as naming and shaming 'Mountebanks among forensic scientists' in a multi-volume forensic science textbook edited by Richard Saferstein. Starrs lists a number of these 'experts' in forensic matters and characterises much of their work as 'scientific hokum'.

Indeed, we have had a parade of definitive pronouncements from 'experts' who are not. Where this occurs within the legal system, the damage done both to people's lives and to justice is incalculable. During my years of practice, I have seen overstated forensic evidence, state experts who simply would never make any concessions, judicial officials who could not understand why a state official would exhibit such bias as he or she did, and a growing number of prosecutors who indulged in devious and unethical behaviour.

Forensic science is a very eclectic subject. It is impossible to investigate any case without crossing various boundaries between subjects. It is, I have found, an amalgam of biology, chemistry, mathematics, physics, anatomy, microbiology, statistics, pathology and the law. Often one has need of specialist help in these fields, and I have never hesitated to call in such help from the specialists. However, unless you have some understanding of these subjects, the idea of calling in help may never occur to you.

I have always been a firm proponent of the empirical approach. There are often so many variables that to rely on past experience can produce what is no more than anecdotal evidence. In my opinion, the most dangerous words in medicine and forensic science are 'in my experience'. This is not a phrase you will hear from a chemist or a physicist. Judges should learn to distrust evidence based on experience at first sight. Such evidence, unless backed up by hard data, is useless, dangerous and prone to be misleading. This type of statement is often to be heard from forensic pathologists and forensic scientists. All too often, the presiding officer in a court will be asked to place great store on the number of years a pathologist has practised or the number of post-mortems he or she has done. One good and accurately documented post-mortem, however, is worth more than 10 000 badly performed ones.

My early career in forensic science was a new experience for me, one for which academic life had not fully prepared me. True, I had lots of book learning in a variety of areas, and what I did not know I knew where to find. The various fields to which I had been exposed as a student – biochemistry, physics, pathology, serology, surgery, anatomy – proved invaluable in the world of forensic science. Yet my experience of the law and its interaction with forensic science was a new-out-of-the-box shock.

Because I was literally the only private forensic scientist in the country at the beginning of my career, work came from many quarters. The insurance industry started to use me quite early on. It became apparent that, although this could be a lucrative avenue of work, they wanted exclusive access to my services, which I quickly realised would be fatal to my independence. Within my first few cases with one of the insurance giants, I had their claims manager telling me what he wanted in my report. I refused, and the work from this insurance company dried up in an instant. Most of the others also demanded absolute loyalty. In a particularly foolish moment, one of the senior men at another insurance giant said to me, 'If you work against us, you will never again work for us.' The fellow would presumably have wanted me to go into a witness box to testify on their behalf and proclaim under oath my independence. Although they never asked me outright to 'doctor' my evidence, the subtle and sometimes not-so-subtle pressure was on to find what they wanted.

In the world of criminal forensic science, I bumped into the infamous Lothar Neethling. In the beginning, he instructed his staff at the laboratory to have nothing whatsoever to do with me. In a case early on in my practice, I was estimating the distance at which a shot was fired from a shotgun. The spread of the multiple pellets of shot on a target enables one to calculate the distance at which the shot was fired. However, as with all scientific measurements, there is an inherent variability that has to be factored into the calculation using statistical methods. During an encounter with the then head of the police ballistics unit, I asked him if he applied statistics to such measurements. 'Yes,' he said, 'we keep a record of each case.' There was no point in continuing the conversation. I also became aware of the strong random correlation of striations on bullets from different weapons and, in a paper in around 1995, I wrote: 'It is clear from the literature that random matches of striae are more common than is generally known. The mechanism whereby this phenomenon occurs has been related to the nature of the grinding and sharpening equipment used in the production of firearms ... The subjective nature of the evidence renders it open to abuse, particularly by scientifically untrained police ballistics experts.'

I presented this evidence before the late Judge Schutz. Unfortunately, this set of concepts may have been before its time and the evidence did not find traction in that court. It was unfortunate, because the issues I raised at the time have now, some twenty-five years later, become a major topic of contention in the new scientific paradigm that is changing the face of forensic science.

This book tells the story of some of the so-called experts who have paraded themselves through the courts, and gives an account of the terrible miscarriages of justice that have often resulted. Unfortunately, we are not unique in South Africa. The maladies that I chronicle occur throughout the world, across legal systems. Never has there been a greater need to examine what the state offers as forensic evidence, and never has there been a greater need for scepticism and vigilance.

The necessity for watchfulness in the world of prosecutions and forensic science is amply demonstrated by the headline in the *Cape Times* of 28 March 2014: 'Man freed after 45 years on death row'. Nowhere is the ultra-reliance on the extraction of confessions more starkly illustrated as a faulty method for solving cases than in this matter. Mr Iwao Hakamada spent forty-five years on death row. Only because of the extreme inefficiency of the Japanese justice system was he not executed. His conviction followed what was, it seems, the faulty extraction of a confession from him using methods that we now know are highly likely to produce false confessions. The ability of modern forensic technology to exonerate him must stand as one of the more significant milestones in the onward march of forensic science.

Chapter 1

SIR BERNARD SPILSBURY: 'THAT INCOMPARABLE WITNESS'

'The three most dangerous words in medicine are, "In my experience ..."'

– Mark Crislip

'All men make mistakes, but a good man yields when he knows his course is wrong, and repairs the evil. The only crime is pride.'

– Sophocles

When Isabella Banks took ill and died in March 1859, it was suspected by the medical practitioners who attended her final illness that she had been given some form of irritant poison. Isabella's remains were examined and analysed by Dr Alfred Swaine Taylor, who found arsenic in her tissues. He also detected arsenic in a bottle belonging to a man by the name of Dr Thomas Smethurst, who was bigamously attached to Isabella Banks. The police charged

Smethurst with murder, and shortly afterwards he appeared at the Central Criminal Court – commonly known as the 'Old Bailey' – in London. But by the time Taylor testified for the Crown, he had reversed his views.

In his analysis Taylor had used the Reinsch test, which involves heating a copper strip in a solution produced by grinding up the body tissue in hydrochloric acid. If a silvery coating appears on the copper, mercury may be present, while the appearance of a dark coating is indicative of the presence of arsenic or certain other metals. In Taylor's analysis, a dark coating had appeared. The toxicologist had made the fundamental error, however, of not performing a control test, which would have involved doing the whole test without using the actual body sample. Had he done so, he would have quickly seen that the arsenic he detected came from impurities in the copper strip that he used. The arsenic detected in his sampling of the bottle in Smethurst's possession could also have been released from the copper by the potassium chlorate present in the bottle.

This was an elementary mistake and, to his great credit, Taylor informed the court at the murder trial of his conclusions, admitting that the arsenic he had measured came from imperfections in his apparatus. Despite Taylor's admission and the fact that two other witnesses for the prosecution contradicted each other, Smethurst was, inexplicably, still found guilty and sentenced to death. Later on, good sense prevailed and the conviction and sentence were set aside by Sir George Cornewall Lewis. (Interestingly, Sir Samuel Wilks, who carried out the post-mortem on the unfortunate Isabella, found severe colonic ulceration. He published his findings in a letter to the *Medical Times and Gazette* in 1859. It would seem that Isabella may have died of an ulcerative colonic disease now known as Crohn's disease, which is still not fully understood.)

Poor Dr Taylor was subsequently criticised in the *Dublin Journal of Medical and Chemical Science* for having brought 'an amount of disrepute on his profession that years will not remove'. Taylor may have made an awful mistake, but he was man enough to reverse his opinions, at great personal and reputational cost. The debacle certainly set forensic science and forensic medicine back many years.

Forty years later, in 1899, there entered this scene of forensic medicine a man who, despite modest beginnings, was to become a household name in forensic medicine in England: Bernard Spilsbury.

The 1951 biography written by Douglas G. Browne and E.V. Tullett, *Bernard Spilsbury: His Life and Cases*, describes the young Spilsbury in unflattering terms. One of his Oxford contemporaries allegedly said of him, 'I regarded him as a nice, very ordinary individual and certainly never expected him to do anything brilliant.' In his formative years, Spilsbury fell under the influence of Drs A.P. Luff, William Wilcox and A.J. Pepper. Under the tutelage of these three, he entered the field of forensic medicine. Two years were to pass before an important case propelled Spilsbury to the front line of his profession, and into the public eye. The case centred on a man called Hawley Harvey Crippen.

Crippen was a diminutive American with dubious medical qualifications from the Cleveland Homeopathic Hospital in Ohio. He also had diploma qualifications from the Ophthalmic Hospital in New York, which enabled him to practise as an eye and ear specialist.

After the death of his first wife in 1891, he married again. Cora Turner, whose birth name was Kunigunde Mackamotski, was an American of German and Polish-Russian extraction. Crippen was to have a fairly unspectacular professional career in the United States, and moved to England in 1900 with his new wife. During his partnership in a company known as the Yale Tooth Specialists, he employed as his typist a woman named Ethel Le Neve.

Some time in 1910, Cora Crippen went missing. After a while, her friends became concerned and contacted the police. In the meantime, Crippen was seen out and about with Miss Le Neve, and it was known that he had started to pawn his wife's belongings. Police inquiries were carried out by Chief Inspector Dew, to whom Crippen gave some anodyne story about Cora having left him to return to Chicago. Dew was not convinced by Crippen's account. A day or two later, the police inspector returned to the Crippen household to confirm a couple of dates, only to find that Crippen had flown the coop, accompanied by Miss Le Neve.

Now thoroughly suspicious, Dew went through the Crippen house with a fine-tooth comb. In the cellar, buried below the brick floor, he discovered dismembered human remains. The hunt was now on for Crippen and Le Neve.

The pair were on their way to Canada aboard the SS *Montrose*. Le Neve was disguised as a boy. The captain of the vessel, Captain Henry Kendall, became very suspicious of the affectionate behaviour between the odd couple and made use of his then new-fangled radio to alert the British police. Chief Inspector Dew immediately boarded a faster vessel, the SS *Laurentic*, and overtook the *Montrose* in time to arrest Crippen and Le Neve before they could disembark at the mouth of the St Lawrence River in Quebec.

The trial was a high-profile one. Apart from the fact that this was the first time that ship-to-shore wireless telegraphy had been used to apprehend a suspect, the sensationalist details consumed the tabloids. After five days, Crippen was convicted and sentenced to hang. In a separate trial, Ethel Le Neve was defended by F.E. Smith (later Lord Birkenhead, who became a prominent politician and a close friend of Winston Churchill) and was acquitted.

Key to the prosecution in Crippen's trial was proving that the remains found in the cellar were those of Cora Crippen. Bernard Spilsbury's evidence was devastating and decisive; it was arguably the most damning of the evidence that linked the remains to Cora. Spilsbury, who had recovered the remains, had made microscope slides of them, one of which showed a piece of skin with a marking allegedly consistent with an abdominal-surgery scar of Cora's. In the witness box Spilsbury was unshakeable and resolute, and his evidence played a major role in convincing the jury that the remains belonged to Cora Crippen. The presiding judge, Lord Richard Alverston, stated to the jury: 'Gentlemen, I think I may pass for the purpose of your consideration from the question of whether it was a man or [a] woman. Of course, if it was a man, again the defendant is entitled to walk out of that dock.' It was of course impossible at that time to determine the sex of the remains scientifically, and the court believed the evidence Spilsbury presented.

Hawley Harvey Crippen was hanged at Pentonville Prison on 23 November 1910. Throughout the trial, he had proclaimed his innocence, declaring just before his hanging, 'I insist I am innocent ... some day evidence will be discovered to prove it.'

After a century, Spilsbury's slides were rediscovered. They were subjected to scrutiny using modern forensic DNA-analysis techniques. The thin slice of tissue that had been mounted on the crucial slide was removed and the DNA was extracted. Dr David Foran and his co-workers, who conducted the research, published their findings in the *Journal of Forensic Sciences* under the title 'The Conviction of Dr Crippen: New Forensic Findings in a Century-Old Murder'. What they found was astounding: 'Based on the genealogical and mitochondrial DNA research, the tissue on the pathology slide used to convict Dr Crippen was not that of Cora Crippen. Moreover, the tissue was male in origin.' These scientists had traced Cora's living relations and had been able to show that the mitochondrial DNA, which is passed exclusively down the female line, was completely unrelated to her. Proving that the tissue was male put the clincher to the case.

So where did it all go wrong? Here we had Spilsbury, red carnation in buttonhole and absolute in his beliefs. He was so confident and certain that he brooked no doubt – which is just what the judge, who was not able to evaluate the evidence for himself, wanted. Many legal men, even today, are scientifically illiterate. There was only one problem: Spilsbury was dead wrong.

This matter might be no more than a footnote were it the only case where Spilsbury's dogmatic confidence resulted in a wrongful conviction and, indeed, a wrongful execution. But it is not. The trial of Norman Thorne, which played out amid intense press coverage over five days in April 1925, is another. Playing centre stage was 'that incomparable witness' Bernard Spilsbury.

The facts of the case are as follows: Elsie Cameron, who lived in North London, travelled to the village of Crowborough in East Sussex to visit her fiancé, Norman Thorne, an unsuccessful chicken farmer eking out an existence on a smallholding near the town. After 5 December 1924, Cameron was never again seen alive. Her disappearance became a national preoccupation and, when eyewitness evidence placed her at Thorne's smallholding on the day of her disappearance, the police questioned him. He denied that she had been there. A search of the farm quickly turned up her luggage and Thorne was taken in for questioning. The farm was further searched and, after a few hours, Elsie Cameron's body was found buried under the chicken run.

The couple's engagement was lacklustre. Cameron has been described as 'an unattractive figure – neurotic, not even clever, unable to keep a post because of ill-health or inefficiency, her little mind filled almost exclusively with thoughts of sex and marriage'. Thorne, in the meantime, had become attracted to a local girl called Elizabeth Coldicott.

Of course, when Cameron got to know of the interloper into her imagined world of marriage and security, she was incensed and went to Thorne with a fierce determination never to leave. Nor did she.

Thorne's initial denials that she had ever arrived on the farm were quickly refuted by the findings of her luggage and, shortly afterwards, by her decomposing and dismembered body. Spilsbury was called by the police to perform the autopsy, which he did on 17 January 1925, almost six weeks after her disappearance. Once the body had been found, Thorne had changed his version of the story. His new story went as follows: After a heated altercation about Elizabeth Coldicott and having been backed into a corner by an increasingly hysterical Elsie Cameron, who was claiming to be pregnant and demanding marriage, Thorne left his hut. When he came back two hours later, he found Cameron partially suspended from the ceiling beam in the hut. He immediately cut her down. Thorne claimed that he thought he would in some way be held responsible, so he cut the body into four pieces and buried them in the chicken run.

Spilsbury's autopsy showed a healthy body with no broken skin or bone, no discolouration of the face and no organ damage. However, he found signs of significant physical trauma 'which were not apparent on the surface and could only be appreciated on deeper dissection'.

A second post-mortem was performed on the re-exhumed body on 24 February 1925 by Dr Robert Matthew Brontë, a pathologist from Harrow Hospital and later the Crown pathologist for analysis in Ireland. Since Elsie Cameron died on or around 5 December 1924, Spilsbury's autopsy, which took place on 17 January 1925, was performed some five and a half weeks after her death. Brontë's autopsy took place a further five weeks after that. Decomposition by this stage was significantly advanced.

Thorne's subsequent version of events was always consistent: Elsie Cameron had hanged herself after an argument about marriage. The crucial elements of this case turned on whether she had been savagely beaten, thus producing the bruises that Spilsbury found at the initial post-mortem, or whether there were signs of a ligature around her neck.

Brontë and another pathologist, Dr Hugh Millar Galt, gave evidence for the defence and could not reconcile Spilsbury's 'savage beating with an Indian club' with the absence of bone fractures and a lack of disruption of the skin. Spilsbury, on examining Cameron's neck, neglected to take tissue samples for histology and relied on observations at the post-mortem alone. The defence maintained that whatever 'bruises' Cameron had were caused when she fell to the floor after being cut down by Thorne. They also argued that Spilsbury had considerably overstated the severity of this bruising.

When Brontë examined the slides of the neck, he contested that there were clear signs of extravasations of blood, or bruising. This would be in keeping with the version proffered by Thorne that Cameron had hanged herself. Spilsbury countered this by claiming that the signs of bruising seen by Brontë were post-mortem artefacts produced during the process of decomposition of the body.

It is worth taking a brief detour to understand exactly what a bruise is and how it may be formed. Normally blood is contained within the blood vessels. Arteries leave the heart and branch progressively, becoming smaller and smaller until they form capillaries. The width, or diameter, of a capillary is just large enough to allow the blood cells to pass through in single file. Thus, we are dealing with minute and very fragile tubes. The capillaries deliver oxygen to the tissues via the capillary bed. Once the blood passes through the capillary bed, it gets taken up in minute venules – small veins that are much the same size as the capillaries – and these join together like the tributaries of a river to form larger and larger veins, which deliver the deoxygenated blood back to the heart. The whole system is under pressure of about 120 to 135 millimetres of mercury and it is the pressure, caused by the pumping of the heart, that causes the blood to circulate. If any injury causes the small capillaries to break, then blood is no longer confined to the inside of the blood vessel and it leaks out into the surrounding tissue. It is this leakage of blood into the extravascular space in the surrounding tissue that is known as a bruise. Microscopically, the blood cells can be seen outside the confines of the capillaries, in the matrix of the tissue.

There was an undoubted clash of opinion in the Thorne trial between the prosecution team, represented by Spilsbury, and the defence team, headed by Brontë and two other experienced pathologists. Spilsbury had failed to examine the neck and sections of the tissue using a microscope and, when confronted with Brontë's evidence of bruising, he simply dismissed it as a post-mortem artefact. When challenged about the small amount of blood seen in the other so-called bruises, he dismissed this again as a post-mortem artefact. Spilsbury maintained that the blood he had observed in his original post-mortem had been 'washed out' by the water in the coffin. In the light of his evidence of physical trauma that could 'only be appreciated on deeper dissection', Spilsbury's dogmatic stance on the issue is inexplicable.

This brings us to an important consideration. Spilsbury was central to a culture that was used to monopolising the entire process of forensic fact-making. It is the prosecution and their forensic experts who are usually first on the scene of a crime (other than the criminal, of course). They gather evidence and, in so doing, often alter the scene irrevocably, which puts the defence at an immediate and irremediable disadvantage. In addition, the vast majority of post-mortems are conducted by the state pathologist, and this again places the defence at a disadvantage, because a second post-mortem is seldom as revealing as the first.

The judge in the Thorne matter, Mr Justice Finlay, did not help the cause of justice by repeatedly referring to Spilsbury as 'an eminent pathologist' while failing to comment on the academic standing of the defence pathologists. Undoubtedly this would have influenced the jury's decision, which was reached after less than thirty minutes. It would have been impossible for the twelve laymen in the jury to have evaluated the relative merits of the opposing views; they did not even make an attempt to view the slides. In the final event, what the jury did was side with the known expert. They accepted Spilsbury's view that bruising could not have occurred after death in the way it was alleged to have done by Thorne's defence counsel. This they did because of his eminence rather than for any scientifically justifiable or rational reason.

Keith Simpson, who rose to prominence at the tail end of Spilsbury's career as a forensic pathologist, had this to say about post-mortem bruising:

> As regards bruising after death, there can be no doubt that it is possible. Heavy blunt injury can tear dead vessels and open up

tissue spaces into which blood may seep passively. Such extravasation of blood will not extend far and the difficulty of distinguishing them from antemortem bruises can be resolved by a little common sense: there is never any cellular reaction.

In this last phrase, Simpson is referring to the cells of the immune system, namely lymphocytes and macrophages (scavenger cells), which are attracted to the site of an injury and are quite easily seen microscopically as having ingested the various bits of debris in the area. Simpson goes on to say:

> It is fair to add, nevertheless, that when injuries occur closely at or about the time of death it may be impossible to say whether they occurred just before, at, or about the time of death ... The blood remains fluid for some time – indeed may never clot at all, and may percolate into spaces opened up by injury at, or after, death.

Sir Sydney Smith, writing some nine years later in his textbook *Forensic Medicine: A Textbook for Students and Practitioners*, is unequivocal. He opines:

> The question is often asked whether it is possible to cause bruises on a dead body. I am of the opinion that although the application of a good deal of violence to a dead body may produce a small extravasation of blood in an area of hypostasis, yet there is no similarity whatsoever between that and an ante mortem bruise ... In undecomposed bodies the distinction between ante mortem and post mortem bruises is easy. The former are characterised by swelling, damage to the epithelium, extravasation, coagulation and infiltration of the tissue with blood; the absence of all of these phenomena from the latter leaves no room for error ... In putrefied bodies where the tissues are discoloured with transfused blood

> pigment, moist with fluid and distended with gases, the diagnosis of small bruises may be absolutely impossible and the observer should never give a definite opinion unless he can demonstrate a local effusion of blood or see some vital reaction in microscopic preparation.

At this point one can start to appreciate the differences in opinion between eminent individuals. There is no doubt that each of them would have advanced their views in evidence with the authority of holy writ.

Turning now to a modern textbook on forensic medicine, Werner U. Spitz and Russell S. Fischer state: 'Disproportionately extensive bruises on the face are sometimes seen in partially suspended victims of hanging after minor trauma. Such trauma may be sustained just prior to hanging or when a suspended limp body swings against a wall or falls to the ground while being cut down.' Even more recently, in 2001, there appeared a paper by P. Vanezis on the difficulty of interpreting bruises at autopsy.

This brief overview of the developments in thinking on post-mortem bruising shows a number of things. It was known at the time of the Thorne trial that the blood in a cadaver can take some time to coagulate, if ever. Spilsbury must have autopsied many bodies where the blood was still fluid hours, if not days, after death. He must have been aware of the basic mechanism of bruising, namely damage to the microvasculature in the area of trauma. In addition, he must have had some knowledge of hypostasis, the process where blood percolates downwards as a result of gravity, producing a mottled appearance in the cadaver's lower areas. It would have required no great understanding to see how the gravitational pooling of blood could leak out of the vessels in the area of trauma to produce a lesion indistinguishable from a bruise. Vital reactions – that is, the body's own immune response to trauma and infection – take time, hours, at least, and so bruising in the immediate post-mortem period, like bruising in the immediate ante-mortem period, would not show any vital signs: the bruising would be indistinguishable.

Spilsbury's dogmatism, coupled with the bias of the court, sealed Thorne's fate, and he was hanged protesting his innocence to the last. The fact that experienced pathologists for the defence gave views that were starkly opposed to those of Spilsbury, and the fact that Spilsbury had not produced or examined microscopic sections of the neck, were to no avail. His position as pathologist for the prosecution, and his fame and demeanour in all likelihood were pivotal in sending Thorne to the gallows. Thorne was certainly foolish to have dismembered the body and attempted to conceal it, but his version of Elsie Cameron's death was more than possibly true, and he should never have been hanged.

Of more than passing interest is that one of the concerned citizens who joined the public outcry against Thorne's conviction was none other than Sir Arthur Conan Doyle, the creator of Sherlock Holmes, the world's most celebrated fictional sleuth. It is also fascinating that in Doyle's 1888 novel *A Study in Scarlet*, Watson meets up with an old friend, Stamford, who tells him about Sherlock Holmes. 'He appears to have a passion for definite and exact knowledge,' says Stamford of Holmes. 'Very right too,' replies Watson, and Stamford continues: 'Yes, but it may be pushed to excess. When it comes to beating the subjects in the dissecting-rooms with a stick, it is certainly taking rather a bizarre shape.' It is spellbinding, and uncanny, that art so pre-empted reality.

Towards the end of his career, Spilsbury played a significant role in another high-profile case, this time an interesting and gripping tale of spies, subterfuge and deceit. The story centres on the planning for the invasion of Italy during World War II, when the Allies were in occupation of North Africa. The invasion across the Mediterranean would involve either Sicily or Sardinia, with the obvious favourite being Sicily. Ewen Montagu, a commanding officer in British Naval Intelligence, dreamt up a plot to make the Germans believe that the invasion would take place via Sardinia and so divert troops away from Sicily, where the true invasion would occur. Codenamed 'Operation Mincemeat', the scheme involved floating a corpse ashore off the coast of Spain. The corpse would be in possession of secret documents revealing that the invasion would occur via Sardinia.

To this end, the British obtained the corpse of an unclaimed suicide by the name of Glyndwr Michael, who had taken a dose of phosphorous-containing poison. In their investigations into the suitability of this particular corpse for the purpose they had in mind, they consulted Spilsbury at his club, the Junior Carlton Club.

According to Ben Macintyre, author of *Operation Mincemeat: The True Spy Story that Changed the Course of World War II*, 'Sir Bernard's verdict was as dry as his sherry: "You have nothing to fear from a Spanish postmortem; to detect that this young man had not died after an aircraft had been lost at sea would need a pathologist of my experience – and there aren't any in Spain."' Spilsbury's opinion, which was accepted by Montagu and his team, was offered with authority and confidence and, like much of his work, was wrong in almost every respect. It would have been simple for any reasonably trained pathologist to detect signs of phosphorous poisoning. Additionally, the absence of indications of drowning should have alerted the authorities. Finally, there were pathologists in Spain who, had they been consulted, would have picked up on the fraud – probably faster than Spilsbury himself. Fortunately, the Germans wanted to believe the story. Had they not been so gullible, Macintyre writes in *Operation Mincemeat*, 'the victims of Spilsburyism could have numbered in their thousands'.

A final word on Spilsbury should rest with the editorial written in the *South African Medical Journal* on 3 May 1952. Titled 'Sir Bernard Spilsbury – a great man, a great witness or a great myth?', the article is largely a comment on Browne and Tullett's biography. The author of this article was the incomparable gentleman of South African forensic medicine, Hillel Abbe Shapiro. With more gentleness than

Spilsbury deserves, Shapiro details one of Spilsbury's cases, *R* v. *Seddon*, where the English forensic pathologist conceded that death could have been due to 'epidemic diarrhoea', but excluded this as the cause because of the extensive preservation of the body on exhumation, which he ascribed to the ingestion of arsenic.

We now know that the amount of arsenic required to retard putrefaction is far in excess of anything that could conceivably have been administered as a fatal dose. We also know that the state of preservation of the body has everything to do with the physical condition in the grave, such as soil moisture content and drainage. 'That the body can be preserved as a result of a fatal dose of arsenic,' writes Shapiro later in *Medico-Legal Mythology and Other Forensic Contributions*, 'is probably one of the greatest myths that pervade forensic medicine.' He goes on to note, 'Some may go so far as to say that Spilsbury's so-called infallibility was no more than a hypertrophic forensic effrontery.' This is a view with which I concur. Modern research has exposed just how wrong Spilsbury could be.

In the Thorne matter, the fact that three well-qualified experts formed opinions counter to Spilsbury's should have been a warning to the court that there must have been some doubt. The comments made during the trial by the judge exposed his pro-prosecution bias, and the failure of Spilsbury to dissect the neck and examine it microscopically was a crass blunder. When other evidence produced at the second post-mortem indicated that Elsie Cameron had probably been suspended by a ligature prior to her death, corroborating Thorne's version, the court should have seen flashing warning lights.

The function of an expert in a court of law is to lead the court along the path of logic and to allow the court to come to its own conclusions. It is not for the expert to usurp the role of the judge and to come to conclusions on the ultimate issue; this decision remains with the court and the court alone.

In Thorne's case, neither the judge nor the jury examined the slides that were so crucial to the outcome. The jury could never have applied their minds to the relative value of the contradictory evidence before them in the short time it took them to reach their verdict. They judged the witness and not the evidence, and that was, and remains, wrong. Today the greatest insult that you can bestow on a modern forensic practitioner is to call him or her a 'veritable Spilsbury'.

The cases discussed above, however, are old. The question to be asked is: 'Has anything changed?'

Chapter 2

'DON'T SPEAK TO THAT BASTARD; THE STATE IS GUNNING FOR HIM'

'I don't want yes-men around me. I want everyone to tell the truth, even if it costs them their jobs.'

– Samuel Goldwyn

The words in the title to this chapter were spoken by Dennis Kemp, the district surgeon of Johannesburg, to David Berson, a relative newcomer to the field of forensic medicine at the time. Berson and I had known each other for several years. Before he entered forensic medicine, he had been a pathologist at the Pneumoconiosis Research Institute, which was adjacent to the old South African Institute for Medical Research, a magnificent old Herbert Baker–designed building in Johannesburg. The background to this extraordinary conversation had some deep roots.

Some fifty years ago, drunken-driving matters, or cases of driving while intoxicated (DWI), were prosecuted by having a good old-fashioned police officer stand up in court and recite the time-worn mantra, 'He was unsteady on his feet, he had bloodshot eyes, his speech was slurred and he smelled of alcohol.' This was often quite adequate to secure a conviction if the accused did not have legal representation. However, three of the four symptoms referred to above can be easily produced by other conditions, such as a head injury or diabetic illness, or, for that matter, any illness producing diminished cognitive function.

To counter this problem, the state turned to the laboratory. Blood samples were taken and these were analysed. Various limits were set, starting at 0.22 grams of alcohol per 100 millilitres of blood. Over the years this was gradually reduced to 0.08 grams and, more recently, to 0.05 grams, which is the equivalent of about two drinks consumed over an hour-long period. The chemical method for testing alcohol was tedious and slow, and so, in the mid-1960s, the gas chromatograph, an instrument that can measure the level of alcohol in the bloodstream, was introduced to the process to speed up the entire operation.

All alcohol of the ethyl-alcohol variety in commercial circulation is produced by fermentation. That is to say, a good supply of carbohydrate, such as starch and glucose, is allowed to interact with a micro-organism, often yeast, which will use the carbohydrate as a food source. One of the waste products of this metabolic process is alcohol (ethyl alcohol). A similar process can be used for the production of vinegar. (These processes are described in more detail in Chapter 13.)

When taking a blood sample, it is crucial to ensure that the sample remains uncontaminated with any micro-organisms that can convert the blood sugar into ethyl alcohol. Significant interest in this field grew after an inquest in 1975 in England. The inquest was into the death of forty-two passengers and the driver of an underground train that crashed into a dead-end tunnel at Moorgate in Central London.

No satisfactory explanation was found to account for the accident. However, ethyl alcohol was found in the blood taken from the driver of the train. Because of the difficulties of extracting the corpses from the mangled wreckage, it was some five days before the autopsy could be performed on the driver, and significant putrefaction had occurred.

At this time, there was very little information on the production of ethanol during decomposition. The upshot of all of this was an important paper written by Janet Corry, which shows unequivocally that ethanol can be produced by micro-organisms in samples of human tissue, including blood. Corry warns that 'extreme caution should be exercised when assessing the relevance of post-mortem ethanol'.

In 1983, I was in charge of the journal club in the Department of Medical Biochemistry at the University of the Witwatersrand (Wits) in Johannesburg. One of my guest speakers was Hillel Shapiro. He was accompanied by Dorothy Gill, an analyst in the government laboratories who was tasked with blood-alcohol determination. She informed me at this meeting and in subsequent discussions that up to 40 per cent of the samples that she tested were contaminated by viable micro-organisms. This creates a problem: the Road Traffic Ordinance specifically states that when a sample is taken for the purpose of determining alcohol, 'It is presumed until the contrary is proved that the syringe used for obtaining the specimen and the receptacle in which the specimen was placed for dispatch to the analyst, were free of any substance or contamination which could have affected the result of the analysis.' Of course, the moment there are viable organisms in the sample, the presumption is rebutted and the accused is entitled to an acquittal.

In 1984, I was briefed in a drunken-driving matter. Before proceeding I phoned Kevin Atwell, who was the Deputy Attorney-General at the High Court in Johannesburg. Kevin was a thoroughly decent human being and, after listening to my concerns, he referred me to the head of the laboratory in Pretoria so that I could speak to them and make whatever suggestions necessary for improvement. This I did. I was given an unfriendly reception and told that they would contact me when they 'needed me'.

Feeling rebuffed, I again phoned Atwell. I was all along disturbed that people would be acquitted on this technical issue, one that was, moreover, so easy to remedy. Atwell said to me that his hands were tied and that I should go and 'try to win a few cases'. This I proceeded to do. By chance, I met up with an attorney called Nic van Wezel and an advocate named Solly van Nieuwenhuizen, and together we cut a swathe through the prosecution for drunken driving.

The issues became very technical in the end, with the legal debate extending to whether or not the preservative (sodium fluoride) added to all samples for blood-alcohol testing was adequate to prevent the formation of extraneous alcohol. I did experiments at the time and I found that the preservation did not entirely prevent the production of extraneous alcohol. At this time I had already suggested that the state change its way of drawing blood. I advised them to use a vacutainer tube, which would have obviated the problem, but I suppose I was quite naive in thinking that the state would take any sort of advice from me. Unbeknown to me, Dr Neels Viljoen, who headed up the Pretoria laboratory, had done his own tests on the question. His results confirmed my findings, namely that a particular organism, *Candida albicans* (the fungus that causes thrush, a very common disease), was able to produce alcohol in the presence of the preservative added to the blood tubes. The state, however, withheld this evidence in various cases in which I was involved.

What is more, each time I gave evidence in these matters, the state would send a team to court to deal with us. On one occasion, they sent Dr Jurie Nel, the chief state pathologist in Durban. He never did get to testify, but the spectacle of these 'eminent', high-ranking pathologists giving evidence on the biochemistry of ethanol production was laughable, particularly in the light of Viljoen's findings. It is equally laughable that these pathologists were happy to come to court to testify on a subject where their own knowledge consisted of second-year elementary medical biochemistry – their brief and elementary exposure as doctors to biochemistry meant that they were hopelessly under-qualified to comment. They therefore did the only thing they knew how to do: attack me personally. This may explain the remarks by Dennis Kemp referred to earlier: 'Don't speak to that bastard, the state is gunning for him.' As I mentioned in Chapter 1, there was an attitude that the prosecution was entitled to be the prime supplier of forensic evidence. Nothing has changed.

Over the years I have experienced almost nothing but obstructionism from the state, especially state pathologists. When I have been involved in a criminal case, I have gone out of my way to prepare bundles of photographs, where relevant, for the prosecution and for the judge. This compliment has never been returned. Invariably, I get given photocopies of the state photographs, which are usually useless for any form of forensic evaluation and often consist only of a black smudge.

In a recent matter where I was acting for the defence, just such a set of black smudges arrived, along with a post-mortem report, from a highly regarded state pathologist. I phoned her to ask if she had copies of the photographs on her computer, to which she replied in the affirmative. I then asked her to email me a legible copy. The answer was a resounding no. I received a lengthy email relating to Section 20(4) of the Inquest Act. I am not sure what the Inquest Act has to do with supplying clear photographs in a criminal prosecution, but the refusal to provide them (when the illegible ones were already in my possession) flies in the face of the provisions of Section 195 of the Constitution, which requires, among other things, that 'Efficient, economic and effective use of resources must be promoted ... Services must be provided impartially, fairly, equitably and without bias ... [and] Transparency must be fostered by providing the public with timely, accessible and accurate information.'

It is noted that these principles apply to administration in every sphere of government. The failure to provide legible copies of photographs therefore goes against the constitutional rights enshrined in Section 195.

This effectively means that the defence in relevant matters is obliged to apply to the investigating officer and, if this police official obstructs the application, make a full-blown court application. This comes at great expense and is a total waste of the court's time when it has to provide something that the accused is fundamentally entitled to in the first place.

The close relationship of state pathologists with the police is, in my opinion, unhealthy in the extreme. Their refusal to discuss the contentious issues with a defence expert is not only unscientific, but fosters a behaviour that fails to serve justice. This phenomenon is not unique to South Africa: one glaring example of this attitude is the melancholy case of Alan Clift.

Alan Clift was a forensic scientist specialising in serology (the study of plasma serum and other bodily fluids), who worked for the Home Office in England. In the early 1970s, Clift was involved in the forensic investigation of the murder of a woman named Helen Will. He discovered that semen found in Will's vagina contained blood-group A substance. Red cells have on their surface chemicals comprising various sugars, which give rise to the blood groups A, B, AB and O. It was realised quite early in the understanding of serology and blood groups that blood-group substance was not confined to the red cell surface; it could also be found in the bodily fluids of certain individuals, who were known as 'secretors'. Not everybody is a secretor: in some individuals, the blood-group substance is, in fact, confined to the surface of the red cells.

After an investigation, a truck driver called John Preece was arrested and charged with the murder of Helen Will. At the trial, Clift gave evidence that Preece was a blood-group A secretor and that his blood group and secretor status matched the material found in the vagina of Helen Will. What Clift did not tell the court was that Will was *also* a blood-group A secretor, so there was no means of distinguishing Preece's from those of Will. That is to say, Clift's findings had no probative value at all.

This evidence was originally contained in Clift's report, but he had been persuaded to edit out the vital piece of information about Will's blood group and so presented a set of half-truths that gave completely the wrong impression. Had he given the full evidence, it would have demolished the Crown's case against Preece. Instead, Preece was convicted on the basis of this worthless evidence. He went to prison protesting his innocence.

Preece won a retrial in 1981, largely as a result of the discovery of Clift's dishonesty. It appeared that the prosecutors had managed to persuade Clift to edit out the exculpatory information on the basis that he would have supplied the correct answers had the defence asked the right questions. In the event, the defence lawyers were not educated or informed in the science of serology and the 'right' question was never asked. Preece was in prison for eight years.

I am at a loss to understand Clift's behaviour. He must have been aware of the consequences of altering his original report. Yet he did it. He must have realised that he would have been forced to explain why the findings had been left out of the report had the 'right' questions been asked. His failure to include a vital finding can only be seen as a dishonest attempt to assist the prosecution. This is not the function of a forensic expert, yet I have witnessed it happening on a regular basis in South Africa.

A very senior forensic pathologist I spoke to recently kept referring to the prosecutors as his 'principals'. In my view, this pathologist is misguided. His principal is the court. As the above-quoted provisions of the Constitution attest, constitutionally he is bound as a state official to provide the information. What he fails to realise is that he appears in court as the witness of the court. If you ask him he will affirm this, yet he will not give information to the defence and he will not discuss the case with the defence expert or, for that matter, the defence's legal representative. This is, in my view, unhealthy. It often means that the first opportunity the defence has of discussing the matter with him is in the witness box under cross-examination – a situation in which he is least inclined to admit to fault or to make any meaningful concessions.

In thirty years of practice in our courts, I have yet to find a prosecution expert who will make a concession. Think about it: not once in thirty years has a state expert admitted to having possibly been wrong. It seems that we are dealing with a race of geniuses … I think not.

Chapter 3

HOW COOL IS COOL?

'Convictions are more dangerous foes of truth than lies.'

– Friedrich Nietzsche

'The bureaucratic mentality is the only constant in the universe.'

– *Star Trek IV: The Voyage Home*

In 2010, a young man and his wife were living on a wine estate in Franschhoek in the Western Cape. Early one July morning, he left for a trip up the West Coast. He was seen on video security footage leaving the farm at 02:00. Just before 07:00, he was a few hundred kilometres up the coast and, when unable to raise his wife telephonically, he phoned his mother, who lived a few hundred metres away. He asked her to investigate. She arrived at her son's house and found the door open and the dead body of her daughter-in-law lying naked on the floor of the bedroom. The discovery took place at 07:00 and the police, when called, summoned the local state pathologist, who arrived at about midday. The pathologist took the temperature of the body as well as the temperature of the surroundings at 13:15 that day.

Anyone who has watched one of the dozens of American crime-scene reality-TV shows will know that the medical examiner inserts a temperature probe (which is a bit like a long meat thermometer) into the body and pronounces an 'accurate' time of death. Our intrepid state pathologist, no doubt well taught by American TV, did just that. She made use of a device called the Henssge nomogram, which allows one to feed certain information into it and then calculates a probable time of death – or, rather, allows one to read it off from tables.

It should be trite science to anyone who thinks about it that a body will cool down much faster if the outside temperature is low compared to a situation where the surrounding temperature is higher. In this particular instance, the pathologist concerned used the temperature at midday for the purposes of her calculation. What she should have done was to obtain the weather data for the area and then to have averaged out the temperature. This would have given a slightly better estimate (but still not a very accurate one) of the post-mortem interval. The reason for this is that the body cools quickest when its temperature difference to the surroundings is greatest (where the outside temperature is lower than the body temperature). So the rate of cooling, which slows down over time, is very quick just after death compared to the rate ten hours later. Attempting to allow for this with a simple formula falls far short of the ideal. Yet, to fail to use the average temperature over the period shows a fundamental lack of understanding of the physics involved.

Using the flawed understanding of Henssge's method, and failing to take into account the low temperatures in the early hours of the morning at the farm, led the pathologist to calculate a very long period between the time of death and the time at which she made her observations. As a result of this misapplication of the body-cooling data, the state pathologist estimated that the deceased had been dead since before the husband left the farm. Her calculations did not take further factors into account, such as the possibility of a slight breeze through an open window, the fact that the body was naked for most of the time and covered for part of the time, and the fact that it was lying on a cold floor.

Much more egregious was the pathologist's ignorance of the literature and her failure to heed the extensive warnings that this method of examination of post-mortem death interval is very inaccurate at best and can be positively misleading at worst.

When I noticed the error that the pathologist had made, I contacted her and suggested that we meet to discuss the problem. I was met with a flat 'no'. The consequence of this is that the case – which is ongoing – consists only of this faulty piece of evidence, whereby the prosecution wishes to place the husband, now the accused, on the scene at the time of his wife's death. Without this vital piece of evidence from the state pathologist, the case against the accused will evaporate. Instead of conceding the problem and learning from experience, the whole juggernaut of the state continues to roll. Costs for the defence are increasing by the day. When we get back into the hearing, the court will be blocked up as this state pathologist is slowly and laboriously shown to be incompetent. The stress on the accused both financially and emotionally is significant and unnecessary.

The system in civil trials is in many ways much more sensible. If expert evidence is to be given, the opposing parties must give copies of summaries of the expert opinions to each other timeously. The purpose of this is to prevent trial by ambush. In recent years, it has become standard practice for the judge president to refuse to allocate a court to the parties unless a joint expert minute is filed. The opposing experts must meet and decide what they agree on and what they do not agree on, which means that most of the issues can be dispensed with and the court will be called to adjudicate only on matters where there are serious irreconcilable differences. The saving in costs and court time is considerable. Usually experts differ when they are not given all the facts. The joint expert meeting is a good place to set things straight.

The negative attitude of the state runs deep. It involves not only the forensic pathologists, but also the state forensic laboratories, and it includes the police and the prosecutors too.

Many years ago, I was appointed by Graham Edwards from

Webber Wentzel, a prominent firm of attorneys, to investigate the killing of Raymond Kobrin, for the defence. Kobrin had gone missing and was finally located some weeks later in the boot of his car at the airport in Johannesburg. He was dead from a single 0.22-calibre shot in the back of the head. Blood was found in the bedroom of the Kobrin home in Bedford Gardens near Eastgate in Johannesburg, and his wife was charged with the murder.

Naturally, I went to the scene of the fatal shooting at the Kobrin home. There I found the front door open and a Warrant Officer Holmes presiding over the investigation. As I entered the house, Holmes greeted me with 'Wie is jy?' (Who are you?)

'I am Dr Klatzow,' I replied, extending a handshake, which was ignored.

'Ja, ek het van jou gehoor,' said the charming Holmes. (Yes, I have heard of you.)

'Really,' I said. 'Where did you hear of me?'

'By Polisie Kollege. Ons het lesings gehad oor jou.' (At Police College. We had lessons about you.)

'What did they tell you?'

'They said, "If that guy comes onto your scene of crime, you tell him fok-all."'

As I have always said, the only compliments I get from state employees are inadvertent.

The problems with state representatives' views of forensic science can be segregated into several categories. Firstly, as I have mentioned, the state and its minions believe that they have the monopoly on forensic fact-making. To this end, they will exclude defence representatives from access to the crime scene, the exhibits and the witnesses and, in the event of any forensic analysis having been done, they will keep you from their raw data. All of these actions are wrong and all of them prejudice the defence, sometimes irreparably. Not only are they legally and morally wrong, but they fly in the face of good science, which is open about its methods and results.

In my early days as a forensic consultant, I was involved in a major criminal trial that became known as the 'Mandrax factory case'. When I asked the prosecution for access to the exhibits, I was refused. Back to court we went to obtain a court order giving us permission to uplift samples of the 'drugs' seized so that we could analyse them. The fellow in charge of the laboratories in those days was Lothar Paul Neethling, who gave us samples of all the exhibits with unconcealed bad grace. Each sample that I collected was sampled by his lackey, one Lieutenant Twigg, who took a spoonful of each of my samples. (I should have smelt a rat then.) When I got home, I found out that they had laid charges of possession of drugs against me. So childish! Fortunately, the Attorney-General threw out all the charges and refused to prosecute.

Secondly, state prosecutors believe that they occupy the moral high ground and that any challenge to them is from someone hired to do and say anything for the sake of a defence attorney who wants them to do so. Nothing could be further from the truth.

I think that the facts should speak for themselves. All evidence presented in court is subject to cross-examination. If you, as an expert witness, attempt to mislead the court, your career in the witness box will be short and unhappy. I have been a court witness for almost thirty years and I have been frequently cross-examined by some very skilled cross-examiners. I have survived, and that comes from being careful and honest in the witness box. I have also never batted for a particular side and, if concessions have had to be made in the box, then I have made them willingly and politely.

We should examine whether the same is true of state witnesses. The dismal and dishonest display put up by the state witnesses in *State* v. *Van der Vyver* is a case in point. The court found the shoe-print examiner in that matter, Bruce Bartholomew, to have been completely dishonest. As the local shoe-print expert in Fred van der Vyver's prosecution for the killing of Inge Lotz, the Stellenbosch mathematics student, Bartholomew claimed that he had matched a bloody mark in Inge's bathroom to Fred's running shoe. I was shown this so-called match early on in the case (see photograph). My view was that not only was there no match, but it could not even be said that the mark was made by a shoe.

In performing any match between two objects, there are several requirements. Take, for example, a tyre print. Thousands upon thousands of tyres are produced from a single mould. Each will have the same tread pattern. The driver then fits this set of tyres to the car and drives the vehicle. As the tyre is used, it starts to show signs of damage – it may be cut or a nail might embed itself in the tread. This damage will be localised at a particular portion of the tread. It is important to note that, to prevent a tyre from making a vibrating noise at high speeds, tyre-tread designs are irregular, which means that any damage to the tyre will be individualised. It is highly unlikely that the same damage type will occur in the same relative position on two tyres. It is this inherent improbability that enables forensic scientists to perform comparisons and to identify individual tyres. The same general considerations apply to footwear.

The overall pattern on a tyre or shoe belongs to all similar tyres or shoes. This is referred to as a 'class characteristic' – it is shared by all members of that class. The damage or wear pattern is, by contrast, highly individual and is thus known as an 'individual characteristic'. It is this that is so useful in the kind of comparison work performed by forensic scientists. (See Chapter 15 for a discussion of class characteristics in the context of firearms.)

For a comparison to be valid, the class characteristics must match, and then the individual characteristics must match. It matters not a jot if you think there are some matching individual characteristics if the background matter of class characteristics does not match. If, for instance, you think that you have a tyre match, it will be fatal to your case if one tread turns out to be from a Goodyear Radial tyre and the other from a Michelin.

In the case of Fred's alleged shoe print, there were no class characteristics and the individual characteristics existed only in Bartholomew's mind. No other reputable expert anywhere could be found to agree with his match. Bartholomew suggested that the final arbiter of his work should be William Bodziak, who was considered the Delphic Oracle of footwear comparisons. Bodziak was a former Federal Bureau of Investigation (FBI) agent who wrote the definitive book on the subject and who achieved prominence in the O.J. Simpson case by correctly identifying Simpson's shoe print at the murder scene. When Bodziak failed to agree with Bartholomew in the Van der Vyver matter, however, Bartholomew resorted to falsehood. I have learnt in subsequent personal conversations with Bartholomew that he considers Bodziak to have lied in all respects about his findings. I don't think so.

The same can be said for Maritz, the man who testified that a hammer could have inflicted the fatal blows on Inge Lotz's head. He conducted experiments on animal heads and, when the exhibit hammer bent, he used a larger, stronger hammer. He failed to tell the court about this change and prevaricated on cross-examination; the information had to be dragged out of him.

In the civil trial for damages following a malicious prosecution, Gert Saayman, the head of the University of Pretoria Department of Forensic Medicine, was pitted against forensic pathologist Linda Liebenberg in a diametrically opposed set of interpretations of Inge Lotz's wounds. One thing is 100 per cent certain: they cannot both have been right. Another thing is just as certain: had the size of the wounds been matched to the size of the hammer head, at best the match could have shown that the hammer might have inflicted the wounds on Inge, falling far short of the standard of proof required to link the hammer conclusively to the crime.

That state pathologists are not always honest is shown in what we now know of the horrifying atrocities committed by the apartheid-era South African Police (SAP), Vlakplaas, the Civil Cooperation Bureau and hit squads. Their activities produced a host of battered people and mangled corpses. Murdered activists would have passed through the state mortuaries on a regular basis. Signs of torture must have been visible to district surgeons and forensic pathologists in the employ of the state, and the police versions of what had occurred must have clashed with the findings in their post-mortem examinations. Was a voice from this august body ever raised to draw attention to the atrocities? Only one. Wendy Orr spoke out, and in voicing her concerns she was rounded on by her more senior colleagues. Drs Tucker and Lang had learnt nothing from their infamous involvement in the Steve Biko murder, when they allowed a barely conscious, badly beaten Biko to remain chained up, naked, in a cell and failed to report any bodily injuries, despite clear external injuries to the face and head as well as evidence of neurological damage. They offered Orr no support and were, in fact, openly antagonistic towards her. She was left to walk the road alone. The deaths in detention were all covered up by multilayered state machinery that had as its central feature loyalty to the system, however wicked that system was.

The forensic behaviour of the state experts and police in a host of cases in which I was involved in the apartheid years was less than truthful. Yet the various forensic pathologists never broke rank. There were some very good people among them, but they were often inhibited by the hierarchy. Generally, they resented being challenged and took any disagreement as a personal affront.

Recently I was briefed in a matter involving a shooting by a policeman. The policeman had been informed by crime intelligence that present in a particular house was a man who was not only in possession of an illegal firearm, but who was dangerous and would not hesitate to shoot. After some difficulties, the policeman gained entry to the house and saw the suspect moving across a passageway, from one bedroom to another. He could see that the man had a firearm. He called on the suspect to drop the weapon and to give himself up. The suspect appeared from the bedroom, adopted a crouching position and started raising his right arm with the firearm grasped in his hand.

In that split second, the policeman had to evaluate whether the suspect was divesting himself of the pistol or preparing to fire a shot. The police constable thought the latter and fired two shots. The first was a warning shot, and the second struck the suspect on the point of his chin. The bullet travelled downwards and through the common carotid artery, through his shoulder blade and then through a door. The bullet hole in the door was exactly at the same height above the floor as the exit wound in the deceased's back (the deceased being in a crouching position when the bullet hit). The lethal wound was the transection of the right common carotid, which resulted in a fatal bleeding out, or exsanguination.

Yet the blood spatter was exclusively behind the door, so the police sent their blood-spatter expert, one Colonel Kock, to the scene. Owing to the location of the blood spatter, he came to the view that the fatal shot had to have been fired through the door, with the shooter on one side and the suspect on the other.

Two factors are important about the second shot fired by the policeman. Firstly, the shot was not a knock-down shot. The wounded man, although fatally wounded, would have had time to move back and behind the door in front of which he had been shot. Secondly, although a major blood vessel in the neck had been severed by the bullet, external bleeding would not have been immediate. This means that, in the few seconds during which he could still move, and before his blood pressure dropped, the deceased could easily have moved behind the door and collapsed. During and after his collapse it is highly likely that he bled to death behind the door.

Unfortunately, the blood-spatter analysis done by Kock was simplistic in the extreme. His knowledge of anatomy, which might have enabled him to evaluate the bleeding pattern with more accuracy, was simply not there. My attempts to discuss the issue with Kock prior to the trial were unsuccessful. I gave evidence in the matter and

Kock attended throughout my testimony. Not a single piece of my evidence was either challenged or rebuffed effectively. Kock added nothing to the cross-examination. This again contradicts what scientific debate is all about: it is not about making an assertion and sticking to it through thick and thin; it is about argument, in the true sense. There must be some dialectic. With the state experts, I have seen, there is none.

Jeanine Vellema, head of the Department of Forensic Medicine and Pathology at Wits, in personal communication with me, said that she was against any communication between her pathologists (who are state medico-legal experts in most murder cases) and defence experts. The reason advanced is that this would introduce bias into the debate. I fail to see the logic in this statement. If her pathologists are so weak that they cannot debate issues with the defence expert, then they should not be there in the first place. Since the Enlightenment the method of science has been one of confrontation and robust debate. In science nothing is hidden, and it is of cardinal importance that raw data be made available to anyone wanting to examine it.

Vellema is of the view – a widely held one in prosecutorial circles – that defence experts are but 'hired guns' to antagonise the state pathologists. In other words, if you are not with us, you are against us. The view that forensic scientists and pathologists are 'agents' of either the state or the defence is misguided; as stated earlier, they are all agents of the court.

In 1987, I was involved in the investigation of the shooting of Ashley Kriel. Kriel was a twenty-year-old activist who lived in Cape Town. By 1987, he had become a thorn in the flesh of the South African government, and especially the SAP. The reasons for the police's antipathy were not hard to fathom: Kriel was a coloured youth cast in the mould of Steve Biko. He was a successful orator and organiser and, above all, a very effective political activist who was able to drum up significant opposition to the hated apartheid regime. It was therefore natural that the police, who were, after all, the de facto guardians of the apartheid system, would develop a deep hatred for this young man who was causing such upheavals in and around Cape Town.

Information obtained from a police informant disclosed that Kriel was staying in a house in Athlone. On 9 July 1987, Warrant Officer Jeffrey Benzien and Sergeant Anthony Abels went to the house at 8 Albemarle Street disguised as municipal workers. They had been instructed to keep the house under observation and, specifically, not to affect an arrest. According to the police version of events, both men went around to the back of the house. Abels was instructed by Benzien, the more senior of the two, to knock on the back door. Quite why this was done is not clear, considering their orders.

After the second knock, Kriel opened the door and was instantly recognised by both policemen. He was carrying a towel and a jersey over his hands and the policemen immediately suspected that he had concealed a firearm or a hand grenade under these items. Whatever questions were asked obviously alerted Kriel to the fact that something was amiss, and he moved away from the door. At this point, contrary to orders, Benzien decided to arrest Kriel and identified himself as police. According to Benzien's version, Kriel immediately uncovered the pistol and pointed it at Abels. A major wrestling match ensued and Kriel was forced to the ground. Benzien had by now gained control of the pistol and he used it to strike Kriel on the head. The two policemen were under the impression that Kriel was unconscious, so Abels proceeded to handcuff the young activist, who suddenly lunged for the pistol. In the ensuing struggle, Kriel broke loose and made for the back door, but Benzien tackled him. With his left hand around Kriel's waist and the confiscated pistol in his right, in contact with Kriel's right shoulder, the pistol discharged and Kriel was fatally wounded in the back.

Ignoring the inherent improbabilities of this version (taken from the judgment of the case), we have the following 'facts': The pistol was in Benzien's hand. The pistol muzzle was in contact with the right shoulder of the deceased. It discharged during the fight.

I was appointed by the Cape Town firm of attorneys E. Moosa and Associates to investigate. I was informed that they had briefed the well-known advocate Jeremy Gauntlett in the matter, so the 'game was afoot', as Sherlock Holmes would have said.

With some difficulty, I obtained access to the weapon involved, a Star semi-automatic 0.22-calibre pistol. I also obtained the same type of ammunition from the manufacturers, Swartklip Products in Somerset West. At the time, I was given only photographic access to the shirt and tracksuit top worn by the deceased. What struck me immediately was the size of the hole in the clothing. I procured some pigs' heads and obtained material as similar as possible to the material from which Kriel's clothes had been manufactured – a cotton-polyester knitted fabric. After placing the material over the pig, I fired a loose angled contact shot into the pig's head. The test was repeated several times, each time resulting in a massive hole being blasted in the shirt.

When a firearm is discharged, flames and hot gases are fired out of the muzzle of the weapon. In the case of a revolver, flames and hot gases are also blasted out of the cylinder gap. Because of the high pressure and the high temperature of the gas, the material is melted and burnt, and in the process a hole is made in the clothing.

There is no comparison between the small hole in the tracksuit top worn by Kriel at the time of the shooting and the massive size and shape of the hole made in the material during my experiment (see photographs). The bullet hole in Kriel's back indicates that the fatal shot was undoubtedly a contact shot of the loose, slightly angled variety. One can see this from the elliptical shape of the soot mark on its upper side, slightly to the midline of the body. (A right-angled shot, by contrast, would have produced a circular disposition of soot staining.) It is also clear that the lower edge of the square-shaped slide of the weapon made contact with the skin during the discharge of the shot – two linear indentations in the skin caused by this contact are plainly visible.

How do we reconcile the obvious contact wound in Kriel's flesh with the negligible damage to his clothing? In my view it is impossible to reconcile the two. I think that the true events unfolded along the following lines:

The two burly policemen surprised Ashley Kriel and quickly overpowered and handcuffed him. Thereafter, Benzien or his fellow officer shot Kriel. Having done so, he needed a plausible story to avoid another 'Biko affair', so he pulled up the clothing and, carefully angling the weapon so that he could see what was going on, he fired a second shot through the same hole in the flesh.

When the police were made aware of my findings and opinions, they immediately did their own tests on the clothing. The ballistics expert for the police was Warrant Officer Willem Visser from the Central Ballistics Laboratory in Pretoria, a pleasant man who had about four years of practical ballistics experience under his belt at that stage. Visser undertook 'scientific' experiments to refute my evidence. From the outset, his approach was deeply flawed. He said in evidence: 'I did not doubt for one minute that I was right, but I had to make certain.' In a scientific inquiry one should embark on one's quest with an open mind. To approach it like Visser did was to invite observational and conformational bias in as guests. His lack of scientific training was very much in evidence. He went on to say, 'What I was trying to do was to find the right medium to act as a background so that I got the best effect, so that I could show and prove that the shot was a contact shot and not a distant shot.'

During the course of his experiments, Visser fired between 180 and 200 shots. He kept no proper experimental notes and naively told the court that he recorded only those results that 'worked'; in other words, those that agreed with what he wanted to find. This goes further than just scientific illiteracy; it is scientific fraud.

In his quest for 'the right results', he changed the background from pig skin to a large rubber block and, finally, to a sandbag. The comparison between a loose angled contact shot into clothing with skin behind it is in no way comparable to a vertical hard contact shot with a sandbag as the backdrop. This was the only way in which Visser could attain the small hole visible in Kriel's clothing, and has nothing to do with the actual shot. When questioned during the inquest, it was put to him: 'If you used the firm portion of the pig abdomen, did you get the same effect?' (i.e., a small hole in the clothing). He replied, 'I did not.' In further questioning, Visser was asked, 'So it was only with the sandbag that you got the same effect as the evidence note with a contact shot?' Answer: 'That is correct.' Finally, Visser said, 'The clothing always tore open and there was a bigger hole than I wanted.'

The whole exercise shows the scientific ignorance of this witness. First and foremost, he should have investigated what happened when a contact shot or a loose contact shot was fired through clothing with human skin (or the closest thing, which is pig skin) behind it.

Second, once he had used the same pistol, ammunition, type of clothing and skin as those that characterised the Kriel shooting, his experiment should have ended. Instead, he conceded early on in the cross-examination, 'I became aware with the tests that I was doing that the background material had a large influence on what happened to the clothes.' He therefore altered the material until he achieved a result that suited him.

Third, Visser's admission that he noted only those results that he 'wanted to keep' is grossly unscientific and displays a lack of scientific pedigree.

In a hotly disputed case such as this, one would think that some attempt would have been made to evaluate the evidence professionally. On the bench as inquest magistrate was G. Hoffman, and sitting with him was now-retired Professor Theo Schwär, head of forensic medicine at the Stellenbosch medical school. Giving evidence for the state was Professor Gideon Knobel, who was at the time head of the Department of Forensic Medicine at the University of Cape Town (UCT). I sat through the evidence of Willem Visser and gave evidence in the matter myself. Visser's evidence was challenged by neither of these two experienced pathologists.

Professor Knobel used the muzzle of the firearm to press deeply into his forearm and was able to show that it left indentations similar to those in Kriel's back. It is important to remember, however, that Kriel was clothed – the muzzle was separated from the skin by two layers of clothing. The experiment by Knobel showed only that the muzzle must have been in contact with the skin, and had little or no relevance to the issue of how the fatal shot, or shots, were fired. It was clear to me that Knobel was trying to prove that the police version was correct. This is, again, not the right approach to science. Quite apart from getting Kriel's arm into a position where he could have discharged the weapon into his own back, there was the irreconcilable difference in the size of the bullet hole in the clothing from what it was in the tests done by me and, strangely, also by Visser.

In my view, Visser's grotesque attempts to interpret his results to fit the state case demonstrate not only his scientific illiteracy but also his bias. This is understandable, given his lack of scientific training, but what I have never been able to fathom is the fact that Schwär and Knobel did not remark on this parody of fool's science, and never expressed an adverse comment on the nonsense being paraded before them. There are two possible explanations for this. Firstly, they may not have understood the issue or the design of the experiment – which is highly unlikely. Secondly, the state pathologists may not have been prepared to contradict the police, however bizarre the evidence before them was.

Hoffman, having listened to the prosecutorial attempts to equate the hole in the clothing with a contact shot, decided off his own bat to make a finding that the hole was indeed big enough to accommodate a contact shot. His judgment says: 'The entrance wound in the clothing indeed shows uneven torn edges.' Consideration of the photographic differences between the two holes gives the lie to this finding. Furthermore, the presence of circular abrasions around both of Kriel's wrists shows that he had been handcuffed while being manhandled.

One of the contradictions of those fateful years in South Africa was the fact that, in the heart of the so-called liberal universities of UCT and Wits, there existed forensic departments that seemed to me to be acting outside of the liberal spirit of the institutions in which they were housed. It is inexplicable that, during the worst excesses of the killing squads and the brutality of the SAP towards mainly (but not exclusively) the black citizens of South Africa, not a word was uttered by any of the forensic medicine departments, the sole exception being Wendy Orr.

All of this illustrates, in my view, the unhealthy cosiness of the police, the state prosecutors and the forensic pathology service. All that I have written emphasises the need for qualified outsiders to challenge the system and to keep close watch on the state. It comes as no surprise to me that the state resists this, providing no cooperation. The cases discussed above, both old and new, demonstrate that the unquestioned reliance by the state on the evidence of its hired hands is unhealthy and leads to miscarriages of justice.

Bernard Spilsbury may have died seventy years ago, but his spirit can be found in South Africa and elsewhere to this day. The over-selling of forensic evidence is still with us, and it has been only in recent years that a slow trickle of challenges has promised to bring about major change in the way forensic scientists evaluate evidence. It is time, too, that the state disabuses itself of the notion that it has a monopoly on primary forensic fact-generation.

Chapter 4

THE FICKLE FINGERS OF FATE

'The plural of anecdote is not data.'

– Raymond Wolfinger

'The eyes are not responsible when the mind does the seeing.'

– Publilius Syrus

The use of fingerprints can be traced back to a thesis presented at the University of Breslau in 1823 by Jan Evangelista Purkinje, who was at the time a professor of physiology and pathology. The Czech physiologist developed a classification involving nine separate types of fingerprints. Purkinje, of course, left his name indelibly on both the Purkinje cells found in the brain and the Purkinje fibres in the heart.

During a period of high volumes of Chinese immigration to America, Sir Francis Galton, a British anthropologist and a cousin of Charles Darwin, suggested that fingerprints be used to keep track of individuals, but his suggestion was never taken any further. Galton was said to have shown that fingerprints are constant and do not change from cradle to grave. Sir William James Herschel, who was in charge of the Hooghly district in Bengal, India, in the mid-1800s, was the first to use finger impressions, in an attempt to combat rampant identity fraud. His efforts failed to gain traction and his methods were discontinued after he left India in about 1895.

The first real step forward in the use of fingerprints was in a paper read before a meeting of the British Association for the Advancement of Science in Dover in 1899. The paper, titled 'Fingerprints and the Detection of Crime in India', was authored by Mr Edward R. Henry, the commissioner of the metropolitan police force in London at the time. Henry's name is important in the history of fingerprinting, because it was he who devised the Henry Classification System, which is still in use today.

From the early twentieth century, fingerprinting caught on rapidly. As early as 1900 it was applied in Ceylon, Australia and South Africa, where, in shades of things to come, it was 'used for registration of natives' by the pass office. Much on the origins of fingerprinting can be found in Henry's monograph *Classification and Uses of Fingerprints*.

Although Herschel and Henry loom large in the history of fingerprinting, recognition of the uniqueness of fingerprints is visible in such artefacts as Middle Eastern clay pottery dated to before the first century CE, where the potter impressed his finger onto wet clay before firing a piece. Clay finger seals were also used on documents in the reign of Emperor Qin Shi Huang, who was around some 200 years BCE. Certainly fingerprints have been contemplated as a means of identifying people for a very long time.

As David R. Ashbaugh writes in his book *Quantitative-Qualitative Friction Ridge Analysis*, the science is based on three premises:

1. Fingerprints, or friction ridges, develop in the foetus in their definitive form before birth.
2. Friction ridges persist unchanged throughout life, save for alterations caused by disease (wart developments, for example) or physical trauma (burns, scars, etc.).

3. Friction ridges are absolutely unique and are never repeated, which means that they constitute an absolute form of identity.

With regard to the interpretation of fingerprints, there can only ever be three conclusions: the fingerprint can be linked to an individual; the fingerprint cannot be linked to an individual; or there is not enough detail to reach a conclusion.

The deposition of fingerprints onto a surface results from the oily and salt-laden secretions that occur on most human skin. The friction-ridge patterns on the tips and pads of the finger are very distinctive and behave as a stamp, leaving a mirror image of themselves on the surface that is touched. This is known as a 'latent print'.

The print can be recovered from the surface and recorded in a number of ways, which make use of the physics and chemistry of the material in the secretions that form on the surface of the skin. To visualise fingerprints, the following chemicals are used: fingerprint powder, which adheres to the oily parts of sweat; ninhydrin, a chemical that reacts with the amino acids in sweat to give a purple colour; and silver nitrate, which reacts with the inorganic salts in sweat, causing a greyish print development much like the development of a photograph (the chemistry is not too dissimilar either). The exact mechanics and chemistry of fingerprinting is beyond the scope of this book. Suffice it to say, all genuine fingerprints result from the touch of the friction-ridge skin on the surface on which the prints are later found.

Many situations in the real world of fingerprint evidence can lead to major problems in practice, however.

Bias

In the context of forensic science, bias is the subconscious exclusion of observations that are contrary to the hypothesis held by the observer. In other words, it is a subconscious selective gathering of evidence. It differs from fraud in that fraud involves a degree of conscious deception.

Writing in the *Journal of Forensic Identification*, David L. Grieve of the Illinois State Police Forensic Laboratory in Carbondale states, 'Bias is perhaps the most complex influence upon examiner approach and the most difficult to control. The danger of prejudicial statements prior to a comparison that a latent has to be from a particular suspect is well founded.'

Although bias differs from fraud, the two are often interlinked when it comes to giving fingerprint evidence.

Fraud

More insidious and damaging than bias is fraud. It has been said that fingerprints do not lie, but I can tell you from personal experience that fingerprint experts sometimes do. Exposing them is always a challenge.

One might be forgiven for thinking that there is no reason for a fingerprint officer or expert to commit fraud. Well, there is. In any system where promotion is determined by the casework and the number of convictions, there is a very powerful motive to cheat. Where a fingerprint examiner is faced with a pile of dockets containing fingerprints from a crime scene that are faint, smudged, partial and in a host of ways difficult to use, the examiner could either report that there is not enough detail to compare the print adequately, or cheat. Reporting inconclusive or negative results is not helpful to the officer's career path; success is measured (wrongly) in terms of positive identifications and convictions.

Some years ago I was briefed to advise in a criminal case. The police officer concerned was a local fingerprint expert who had found himself on the wrong side of the law. He had been charged with fraud and defeating the ends of justice. His method for coping with his workload was simple: he would remove the clear inked print from the docket and he would dust it with fingerprint powder. Because the fingerprint powder would adhere to the printer's ink on the inked prints, he was able to lift a great legible print that clearly matched the print of the accused (because it was taken from the accused's print). This newly lifted print would have all the details from the back of the old print copied onto it and the forged print would be used in the prosecution. This type of fraud is brimming with benefits if you do not get caught: the print is easy to read and compare, you know that the match will be good, and the comparison is quick.

Inked prints for the docket are produced by folding the fingerprint paper and placing it on the edge of a table, which makes the taking of an inked print so much easier. The accused in the matter forgot that when a print is lifted using one or other sticky medium, the print lifts in addition to some of the background. In the instance of the police officer charged with fraud, the forged prints showed clear evidence of the folds present on the original prints in the docket.

This insidious form of official fraud was relatively common. In the police unit stationed at Paarl in the Western Cape, there were several such instances. Eventually the crooked police fingerprint experts were apprehended and convicted of fraud and defeating the ends of justice. Such are the crude frauds.

The question often asked is, 'Can fingerprints be taken from one surface and transplanted to a more incriminating surface?' John Mortimer, in his enduring television series *Rumpole of the Bailey*, has a less-than-honest detective shake hands with a suspect being held in the cells. Unbeknown to the suspect, the detective was concealing a lump of gelignite, a jelly-like explosive material, in the palm of his hand. The hapless suspect had unwittingly imprinted his fingerprints all over the gelignite.

In April 1988, while on a study tour of the United Kingdom, I met with a retired fingerprint expert called Peter Swann. During conversations with him I was made aware of certain fingerprint evidence and experiments that had been conducted at New Scotland Yard with a view to investigating the feasibility of transplanting and forging fingerprints. The report by the National Conference of Fingerprint Experts Steering Committee concluded:

(a) Under ideal conditions, very good results were obtained on a variety of surfaces with no obvious indication that transfer had been made.
(b) In three transfers from one surface to another, in one instance only were traces of the transfer medium apparent.
(c) Using a negative of a fingerprint impression and a light-sensitive copper laminate, a mould was produced from which a latex cast was made. The resultant impression appeared to be bona fide.

I would have thought that this information would be important general knowledge to all who practise in the field of forensics. Basically, it affirms that fingerprints are easy to transfer and forge. Furthermore, it can be done with little evidence of the forgery. Any defence legal team would want to know that this is a possibility.

Unfortunately, the report was marked strictly confidential and, as far as I am aware, has not seen the light of day, until now. In my view, the interests of justice are better served by examining all the facts in full view of both prosecution and defence rather than having vital information, gathered at tax payers' expense, kept secret from the defence and the courts.

After a slow start, the use of fingerprints to make a formal identification of individuality in criminal matters became well established and has been the cornerstone of many a prosecution over the past century. It has always been difficult to attack fingerprint evidence, for several reasons.

Firstly, the courts are reluctant to believe that fraud has been committed by a police officer. Secondly, around the world the expertise required to evaluate fingerprints is largely confined to the police forces. Thirdly, the evidence is seemingly so simple:

1. Fingerprints are unique.
2. Your fingerprint was found on the scene.
3. You must have been there.

We shall see that all the premises noted above are not as firmly grounded in science as would be desirable.

It would be so much easier for everyone concerned if criminals left complete, legibly rolled prints at the scene of a crime. Unfortunately, many prints are real horrors. They are often smudged and are usually partial; sometimes they are double touched (in other words, two prints overlay each other); and they can be damaged to a greater or lesser degree. The fingerprint alleged to be that of Shirley McKie, for example, was far from ideal. This is the fingerprint that led to a botched prosecution and a massive controversy in the UK.

Chapter 5
THE CASE OF SHIRLEY McKIE

'Here's freedom to them that wad speak,
Here's freedom to them that wad write,
There's nane ever fear'd
that the truth should be heard,
Save they whom the truth would indite.'

– Robert Burns

It is appropriate that I should preface this chapter with a verse penned by the great Scottish poet Robbie Burns. The trial of David Asbury took place in Glasgow in May 1997. It was a sequel to the murder of a middle-aged, retired bank clerk named Marion Ross, who had been brutally killed. Apart from crushed ribs and other injuries, she had a pair of scissors plunged into her neck. Her body was discovered on 6 January 1997.

Police investigations revealed that there was no sign of forced entry, so the police came to the conclusion that the perpetrator was a person known to the deceased. Shortly before her death, Miss Ross had employed contractors to do some renovations to her home in Kilmarnock. One of the contractors was twenty-three-year-old worker David Asbury, who hailed from Kilbirnie in Ayrshire. Suspicion fell upon him.

None of Asbury's fingerprints had been found at the scene of the crime, but when police searched Asbury's home, they found a biscuit tin containing about £1 800. It was alleged that one of the deceased's fingerprints was on this tin. It was also alleged that the money was stolen from the home of the deceased. Furthermore, reports suggested that one of Asbury's fingerprints was allegedly found in the deceased's home on a gift tag attached to an unopened present, although this was never confirmed.

One of the police officers on the scene was Detective Constable Shirley McKie, who was starting her way up the path of promotion in the Strathclyde police department. Ironically, McKie's father, Iain, had been a senior officer at the same police station and had just retired after some thirty years in the force.

The officers from the fingerprint division examined the biscuit tin in Asbury's home and lifted a print, which was later 'identified' as belonging to the deceased. Asbury denied that he had stolen the money in the tin. Nevertheless, on this slender evidence, he was convicted and sent to prison for life.

The first rule taught in Forensics 101 is to keep your hands in your pockets when you get to a crime scene and not to leave any evidence of your having been there. As Locard's exchange principle goes, 'Every contact leaves a trace' – whatever is touched leaves evidence on the object touched, as well as on the person who touched it.

During the extensive investigation of the crime scene, the Strathclyde police had lifted many prints. It is normal practice to identify all the people with a legitimate reason for being on the premises and to eliminate or identify any prints that they may innocently have left there. This narrows down the investigative field. In all, 428 prints were lifted and photographed at the scene of the crime. It fell to the Scottish Criminal Record Office (SCRO) fingerprint experts to identify and record all of these prints. Two officials, namely Hugh McPherson and Alister Geddes, identified a mark on the bathroom door frame, the infamous Mark Y7, as the left thumbprint of Shirley

McKie. This was of considerable embarrassment to the local police force: police officers are not supposed to blunder about the crime scene leaving fingerprints here, there and everywhere.

Even at this early stage of the investigation there was some disagreement between McPherson and Geddes. Geddes found only ten points of similarity between the print and McKie's left thumb, while McPherson found sixteen. McPherson now sought other officers who could find sixteen points of similarity. He consulted with Charles Stewart, Fiona McBride and Anthony McKenna, who duly found the sixteen points required. There was just one small problem: Shirley McKie denied having deposited the print. In fact, she denied ever having set foot in Marion Ross's home. As if it were not enough that she had left her print at the crime scene, she was now unwilling to confess to the blunder in the face of what seemed like incontrovertible evidence. What was a minor embarrassment rapidly escalated into a career-threatening event. Worse was to come.

During the trial of Asbury, the defence solicitor, George More, caught wind of the dispute between McKie and the other fingerprint officers. When McKie gave evidence, she denied under oath that Mark Y7 belonged to her. Irritation on the part of her supervisors now turned to fury. Her senior officers realised that her denial had placed the other fingerprint evidence in jeopardy.

At the end of the Asbury trial, the jury believed that the deceased's fingerprint had been correctly identified on the biscuit tin and convicted Asbury of murder. By implication, because they believed the fingerprinting officers to be correct, McKie's denials were disbelieved. She was arrested in March 1998 for perjury and taken to the police station in Ayr, where she was humiliatingly strip-searched and thrown into a cell. If found guilty, the charge carried an eight-year sentence.

The fight was on. By the time it was over, many careers and reputations would be sullied beyond repair. Shirley McKie, however, would emerge from the fray bloodied but vindicated. Fingerprinting in Scotland would never be the same again.

In preparation for McKie's perjury trial, a number of additional police officers were brought in to strengthen what must have already seemed like a strong case against her. One of these was Robert Mackenzie, who came to the conclusion that Mark Y7 was made up of more than one touch of McKie's left thumb. Mackenzie averred that he had found twenty-one points of similarity between Y7 and the left thumbprint of Shirley McKie.

The SCRO had assembled a high-power team of fingerprint experts, and McKie found herself at the front end of an official juggernaut intent on destroying her. Early on in the preparation for the trial, she had consulted Peter Swann, who had come to the conclusion that Y7 was the left thumbprint of his client. The team against McKie now consisted of, inter alia, Peter Swann, John Berry and Malcolm Graham, as well as the original officers.

Fortunately, McKie persisted in her denials that Y7 was her print. Then, by dint of great good fortune, she happened on the name of Pat Wertheim. Pat has impeccable credentials, not only in the field of fingerprint technology but also in crime-scene investigation. He had attended advanced courses in fingerprinting at the FBI in Quantico and at the time had a career in the taking and analysis of fingerprints spanning twenty years. I know Wertheim personally. He is larger than life and has a heart the size of Texas. It comes as no surprise to me that he agreed to go to England to examine the prints for McKie at no cost.

The stakes were very high. Not only was McKie fighting for her liberty and good name, but the SCRO – and the Scottish justice system – was fighting for its credibility. In the background, doubt was being resurrected about the original fingerprint that had sent Asbury to prison. At the time of Shirley's trial, the four SCRO examiners were themselves on suspension – a precautionary measure during this intensely controversial period.

In the meantime, Shirley's expert team had grown to include not only Wertheim, but David L. Grieve, the Illinois fingerprint examiner; Alan Boyle, a prominent fingerprint expert who worked for Scotland

Yard in the UK (some might rate him as Britain's foremost fingerprint expert); and Arie Zeelenberg, one of the superstars of the fingerprint world. He has been involved in crime-scene investigation for over forty years. He was head of the National Department of Fingerprinting in the Netherlands for twenty-three years and is respected internationally as an expert in the world of fingerprinting. I have found him to be a softly spoken man whose quiet demeanour belies the immense regard in which he is held.

Such was the team that went up against the SCRO officials on behalf of Shirley McKie. All of these experts were of the view that Y7 did not belong to her. Unbeknown to the defence, five SCRO examiners had declined to attribute the disputed print to McKie. In 'More than Zero: Accounting for Error in Latent Fingerprint Identification', Simon A. Cole lists a number of fingerprint misattributions that contrast with the general evidence given by the fingerprint experts in this case. The jury was out for less than an hour and returned a verdict acquitting McKie.

In the photographs taken from the PowerPoint presentation to the court in the McKie matter, the differences between Y7 and McKie's print are quite obvious. The case leaves a disturbing question: How in the world did such high-power experts get it so wrong? There was not even agreement on whether Y7 was made by a left thumb or a right thumb.

The committee of inquiry deliberating the issues raised by the McKie case 'found it disturbing that a considerable dispute should exist between fingerprint experts on an issue that might, to outsiders, appear to be relatively straightforward'. The committee went on to say: 'It is worth noting the polarised nature of the views held on the lower part of the mark. It appears extraordinary to the committee that one expert could find twenty-one characteristics in agreement and none in disagreement in the lower part of the mark while another expert could find not one in agreement and nine in disagreement in the same part of the mark.'

The standard guidelines state that for an identification to be made, there must be sixteen characteristics in agreement, with none in disagreement. In the McKie trial, Swann was a little more equivocal on the matter. He said in evidence:

> In practice ... experts find characteristics that appear to be in disagreement for which there is some explanation. That often happens. The first chart I prepared for Mark Y7 contained twenty-one characters in agreement. There were some at the top that, at that time, I could not explain, although I explained them later when I got a proper rolled impression of Shirley McKie's print. *At that time, I knew that because there were so many characteristics in agreement – twenty-one – it was an identification*, irrespective of what we saw at the top (emphasis added).

Zeelenberg, however, gave the matter a different slant when he said, 'Identification is the establishment by an expert of sufficient coinciding coherent characteristics in sequence – the sequence is what is important – in combination with the detail of the ridges *and the absence of even one single discrepancy. One single discrepancy stops the identification process*' (emphasis added).

The committee hearing this evidence subsequently declared:

> Assessing the overall position taken by the fingerprint experts on Mark Y7 the committee is incredulous that Arie Zeelenberg could find no ridge characteristics in agreement and twenty in disagreement, and John MacLeod [an expert who wrote a report for the Scottish Executive] could find one in agreement and fifteen in disagreement and yet John Barry and Peter Swann could find between them thirty-two in agreement and none in disagreement.

There is clearly something rotten in the state of Denmark when the experts can disagree so fundamentally.

A possible clue may lie in the approach taken by Swann, which raised concerns with the justice committee, a member of which stated:

> From listening to Mr Swann's presentation, it seemed that he was saying that when he could not find the point in Shirley McKie's thumbprint where it existed in the mark, he went looking in other places until he happened to find a point that looked the same. The only way that he could do that was if he moved 66° around the fingerprint. I reject that approach, which is not valid. One starts with the analysis of the crime scene mark. One does not go looking willy-nilly in the ink print to try to find points that look like it somewhere else in the print.

Zeelenberg had this to say about the identification of Y7: 'There is no way for an expert to look at this print with an open mind and a critical professional attitude to make an identification with the comparison print of McKie.' Strong words, indeed.

Completely separate, but in no less emphatic terms, were Wertheim's comments on the SCRO fingerprint officials:

> Any competent person trained in fingerprint comparison even at the most elementary level of training and experience, conducting an analysis of scene of crime mark UCO 1050197 Y7 (Y7) and comparing that mark to the inked left thumbprint of Shirley Jane McKie, should have no trouble in reaching the conclusion that the mark was not made by Shirley Jane McKie, but had to have been left by some other person.

Similarly, John MacLeod's view was powerful and unequivocal: 'It is my opinion that the differences between the characteristics in the Mark Y7 and those in Shirley McKie's left thumb can be clearly seen and that reasonable care could not have been taken during the comparisons that wrongly made this identification.'

Subsequent to her acquittal, McKie and her father took on the establishment and brought a civil action for damages against, among others, the Strathclyde police chief constable, the Joint Police Board of Strathclyde, the Scottish Ministers and the four SCRO officers who had signed the report of 10 April 1997. In a judgment given on 24 December 2003, Lord Wheatley found that the officers were immune from suit, but that 'this immunity did not cover what was done maliciously'. Eventually the case was settled with an award to McKie of £750 000 by the Scottish Ministers without any admission of liability of officialdom.

This case was to have serious consequences downstream, not only for the SCRO, which was thoroughly discredited, but for other cases. Its reach extended across the border into England, where Danny McNamee had been sentenced for conspiring to cause an explosion in Hyde Park, his conviction based on a single thumbprint.

Where did it all go wrong, and how can it be that the foremost experts in the art of fingerprinting could find so little to agree on? In evaluating the reasons behind this catalogue of disasters, one needs to understand the corporate mentality of government agencies worldwide, not only in Scotland. As Noam Chomsky points out, what hierarchical structures such as governments, government departments and corporate businesses have in common is that they are all essentially totalitarian states.

What kind of freedom is there inside a corporation? You take orders from above and you give them to people below you. Herein lies the first problem. In such an institution, from top to bottom, there is never any advantage in admitting that a mistake has been made. Government and its agencies are almost never willing to acknowledge any mistakes, let alone a major blunder such as the McKie case. The only way that individuals and organisations like the SCRO officials and the Strathclyde police could handle this was to deny that they were wrong.

A typical example of this corporate mentality of denial can be seen clearly in the evidence of Fiona McBride, a member of the SCRO. (McBride's full testimony is available online.) Her answers are revealing about her mindset and those of her fellow officers. During her testimony, she fences with counsel about even trivial things, such as whether she received a document from Mr Findlay (the defence counsel) or whether it was handed to her by the court clerk. She consistently denies knowledge of the trouble brewing over Y7, despite the fact that almost everybody else was in a state of urgent concern. The evidence is typical of the dishonest witness who finds him- or herself confronted with unpalatable evidence and so simply refutes that the evidence is correct.

The first level of the problem lies in the individuals who make up the system. For them to survive and prosper in such a system, they have to absorb and internalise the values of the system. When the system places greater store on being right over being honest, the seeds of the problem are sown and will come to fruition, as occurred in the McKie case.

Another problematic factor in this matter was the fact that, when McKie denied having been in the house and leaving the print, colleagues of the original fingerprint officers were sent to check on and re-evaluate the work done. For the evaluators to truly open their minds and independently evaluate Y7, they would have had to consider the possibility that close friends and comrades in the fight against crime might have been grossly negligent or incompetent, or both. The collegiate spirit of loyalty within the force is not conducive either to whistle-blowers or to letting the side down. Recognising that the original evaluators of Y7 did a wholly incompetent job would simply not have been part of the corporate culture.

A letter addressed to the Minister of Justice in the Scottish parliament, Mr J. Wallace, is illustrative of the attitude adopted by the SCRO. The whole of the SCRO found itself in the unenviable position of Shakespeare's Macbeth: 'I am in blood stepped in so far that, should I wade no more, returning were as tedious as go'er'

(III.IV.135–7). The letter to Wallace, dated 26 January 2000, reads as follows:

Dear Sir,

THE CASE OF SHIRLEY MCKIE

The views expressed below are the views of the signatories to this document and are not intended to reflect the views of either the Chief Constable of Lothian and Borders Police or of that organisation.

On Tuesday 18 January 2000, BBC's 'Frontline Scotland' programme broadcast the documentary 'The Finger of Suspicion'. The programme referred to the case of Shirley McKie, a former detective constable in Strathclyde Police, who had appeared in court on a charge of perjury. The charge related to the identification of her finger impression at the scene of a murder, and to the evidence given by her regarding that identification at the murder trial. Evidence for the defence was given by two fingerprint experts from the United States. They stated that the fingerprint in question had not been made by Shirley McKie. A verdict of not guilty was returned by the jury.

Fingerprint experts within the Identification Branch of Lothian and Borders Police (the largest 'force bureau' in Scotland) feel compelled to state their position regarding the contents of the aforementioned programme.

We have, via the 'internet', examined the material provided by Mr Pat Wertheim (one of the defence experts called in the case) and reached the conclusion, along with experts throughout the world, that the crime scene mark in question was not made by Shirley McKie.

Several approaches have been made to SCRO regarding the

release of the evidence relating to the case but to date they have refused to comply with any of these requests.

This stance has unfortunately brought the whole fingerprint system into disrepute.

The existing situation cannot be allowed to prevail.

There is not a problem with the fingerprint system. There does however seem to be a problem within SCRO. Until that problem is seen to be addressed, the credibility of the Scottish Fingerprint Service and of the individual experts within it, will suffer increasing damage. It is for this reason that the current situation must be addressed as a matter of extreme urgency.

On the strength of the evidence available, the position of SCRO is untenable.

At best the apparent 'misidentification' is a display of gross incompetence by not one but several experts within that bureau. At worst it bears all the hallmarks of a conspiracy of a nature unparalleled in the history of fingerprints.

If SCRO maintain their present position, fingerprint evidence in our courts shall be challenged and discredited at every opportunity.

It is therefore imperative that this matter is resolved as soon as possible.

It is our view that the fairest and perhaps the only way to achieve this is to instruct a completely independent and respected bureau (eg Metropolitan Police Fingerprint Bureau) to undertake a full enquiry into all aspects of SCRO's involvement in this case.

Then, and only then, can the justifiable faith that the general public previously had in the fingerprint system be restored.

The letter is signed by fourteen experts of the Lothian and Borders Police Identification Branch.

The type of pressure brought to bear on the whistle-blowers or those who break rank is easy to glean from the letter posted online in

October 1999 by Grieve, who testified against the SCRO in the McKie case:

To the Members of the Scottish Parliament:
I am one of the American fingerprint examiners, who gave evidence in the perjury trial of Shirley McKie during May, 1999. I am writing to you in the hope that this letter will be read by many of you because I believe what I may offer has direct bearing in the issues before you. What began as the local matter of the homicide investigation of Marion Ross has now achieved such international notoriety that law enforcement personnel throughout the world anxiously await your decision. That decision will not be an easy one to determine, for the aftermath of the McKie trial has spawned waves of rhetoric containing accusations, innuendos, excuses, rationalizations and unfortunate hyperbole. In the end, however, someone is right and someone is wrong, and that, you must decide.

My testimony in criminal court is a matter of public record and one in which I provided my qualifications. As I have stated I have been a fingerprint examiner for 37 years. For the last 20 years, I have instructed nearly 100 examiners how to perform this specialized skill properly for the Illinois State Police, the third largest forensic laboratory in the world. As an instructor, I teach those techniques and procedures which enable a competent individual to determine when identity by fingerprint can or cannot be conclusively established. As most of you are fully aware, fingerprint identification is often the most compelling evidence presented in a criminal proceeding. In spite of recent legal challenges to fingerprint methods throughout the world, the public continues to have confidence in the reliability of fingerprint evidence, as well they should. Public acceptance of fingerprint accuracy has been earned and maintained simply because most practitioners understand just how much influence they may have over another's liberty or life. Therefore, I also teach my students about personal integrity and individual responsibility, about what is the basis for certainty as opposed to arrogance, and about the dire consequences which may result from their failure to meet the highest standards of our profession. In short, I teach the concept of justice. I also inform my students they must acknowledge they are human beings who are quite capable of making mistakes. That is why we have put in place thorough quality control measures which are designed to prevent abuse or misuse of public trust. These are not limited to review of identifications made by others, but measures which ensure professional growth.

I also participated in the inquiry conducted by HMIC [Her Majesty's Inspectorate of Constabulary for England and Wales] during which I was interviewed by Scottish police officers who were at all times competent and professional. I have read the full HMIC report on this investigation and concur with all findings. If enacted, the recommendations proposed offered an excellent opportunity to restore the credibility and dignity to SCRO and would greatly assist in repairing SCRO's severely damaged reputation. During this inquiry I was asked what constitutes differing expert opinion. The question was legitimate, but the explanations offered recently are not. There are three possible conclusions to any examination, that is, I know whose fingerprint it is, I know whose fingerprint it isn't or I simply don't know. Experts may vary in knowing and not knowing, but experts cannot disagree in whose fingerprint it is. Expert opinion has inherent limitations. Two physicians may disagree on a diagnosis when considering symptoms, but one will be right and the other wrong. Both will agree whether the patient is alive or dead.

The substance of the ACPOS [Association of Chief Police Officers in Scotland] report defies the HMIC investigation and wishes to dismiss the matter as merely a difference among experts. This is naive, for someone touched the door frame inside the Ross home and the biscuit tin. Either SCRO is right or SCRO is wrong, and world opinion states the SCRO is wrong. Outside SCRO, no examiner of merit has supported the identifications in the Ross homicide. Simply put, the mark found inside the Ross home was not put there by Shirley McKie and the mark recovered on the biscuit tin was not put there by Marion Ross while she lived. Thanks to the internet, these fingerprints have been examined by experts throughout the globe and no one has supported the SCRO's claim. From the perspective of quality management, two errors made by the same four people in one case is an intolerable situation and requires strong remedy. These errors reveal not only technical failure which caused their commission, but in the continuing omission by SCRO that they are, indeed, errors, they offer insight into collective integrity and responsibility in which the public has trust. Denying that errors occurred in the face of overwhelming conflicting opinions is not a statement of confidence in the four examiners, but arrogance. Denial of wrongdoing disregards the dire consequences of their actions and reveals a concept of justice that only despots could admire.

I took no pleasure in stating to the court that an error had been made by SCRO in the McKie case, but as a man guided by principle, I really had no other option. I happen to believe that no society can call itself free without an unwavering devotion to justice, no matter how obscure and elusive that treasure may sometimes be. Justice is the key thread that holds the cloth of freedom together and, as such, it must be fiercely protected by each and every one of us. I came to Scotland to testify because I realized that if justice was denied to Shirley McKie, someone else would be next, perhaps I, perhaps you. After all, Scotland and America have far more in common than they are different. For me, my concept of justice was inherited from my grandfather, a simple man from Fife but as true a Scot as anyone who ever walked the streets of Leven. As a young man, he had seen first hand what happened when justice had become merely lofty platitudes without fair practice for all. Were he still alive, I fear he would believe a similar time is upon you once again, and he would ask you, first and foremost, to display allegiance to the noble words of David Hume. You have before you the issue of whether you, as representatives of the Scottish people, will assist in adding to Hume's great wall of justice or, instead, you will be party to its further destruction. I urge you to honor the good citizens of Scotland.

David L. Grieve

Grieve's letter makes his stance clear, that the attack by the SCRO against Wertheim (which ought to have been on scientific principles only) took the form of a concerted smear campaign, and that the pressure on the SCRO to maintain its image took precedence over the truth.

Another broad area where the Crown and McKie came to grief is the way in which the courts approached the expert evidence in the matter. The role of the expert in various legal systems has been set out with some clarity. In 1953, Lord President Cooper said of experts, 'Their duty is to furnish the judge or jury with the necessary scientific criteria for testing the accuracy of their conclusions, so as to enable the judge or jury to form their own independent judgement by the application of these criteria to the facts proved in evidence.'

In terms of witnesses, the expert is different in at least two respects. Firstly, he is entitled to make use of other people's work. Normally this would be considered hearsay evidence, but, in the case of an expert, the received wisdom underpinning all scientific inquiry must be taken into account. If it were not so, the scientific witness would be forced to prove all scientific facts back to the dawn of time. Clearly the facts relied on must be universally accepted throughout the scientific community. Crackpot evidence and junk science are not acceptable.

Secondly, an expert witness is different in that he or she is entitled

– and indeed encouraged – to give opinion evidence provided that the opinion is based on sufficient scientific data. The nature, extent and limitations of expert evidence are set out in almost all books on evidence. *Phipson on Evidence* states that 'experts give evidence and do not decide the issue'. In certain extreme situations, it is possible for expert evidence to be dispositive of the case, as in instances where there is evidence that only an expert could provide and it is unchallenged by any other expert evidence.

In thirty years of court work as an expert witness, I have yet to come across this exact situation. There is no clarity in the McKie case as to whether the court actually looked at the fingerprints and could see the different features. It would seem that the court and the jury did not apply their minds to the actual images. The net result of all of this was that the verdict in favour of McKie caused immense embarrassment to the Scottish authorities and resulted in the fingerprint evidence in Asbury's case being revisited. The ultimate consequence was that Asbury's conviction and sentence were overturned.

Much more sinister was the attempt by the investigators in the Lockerbie air disaster to pressurise not only the SCRO but also Grieve and Wertheim. It was revealed in 2006 that the Pan Am investigator, Juval Aviv, told authorities that FBI agents had travelled to Scotland to pressure the SCRO to ensure that the McKie affair was 'swept under the carpet' to avoid any embarrassment in the run-up to the Lockerbie trial. In the same interview, Aviv said that two senior members of staff in the SCRO fingerprint laboratory in 1999 or 2000 had told him that they had misgivings over the evidence against McKie, but had been urged to 'fall in line'.

The trail of injustice that follows from mistaken, botched and fraudulent fingerprint misidentifications is long and devastating, not only to the people falsely accused and, in some cases, convicted, but to the subject of fingerprinting itself. The general public has been fed a diet of lies concerning the absolute authority of fingerprinting for the purposes of identification. In one of the rare instances where fingerprinting was put to the test, seven fingerprints were sent to various laboratories. Only 44 per cent of the fingerprint examiners tested were able to correctly identify all seven against a national database; 56 per cent got at least one wrong and 4 per cent were able to identify none. Not good for a field of expertise that claims papal infallibility. The take-home message is that 22 per cent of the time, wrong evidence of identification would have been given, evidence that would have been difficult and costly to challenge.

It may come as something of a surprise that most of the laboratories in the US, including the fabled FBI laboratory, are not accredited and have actively resisted external inspection and testing. I suspect that the reason for this coyness over testing lies in the hierarchical control that bureaucracy has over science. Image trumps rigour in science. The laboratory staff have loyalty to the FBI first and to science second, if at all. Many exposés of the failure of the FBI laboratories and the organisation in general are now making their way to the shelves.

The Secrets of the FBI by Ronald Kessler is one such. In its early days, the FBI under J. Edgar Hoover developed into the largest and most sophisticated blackmailing organisation in the world. Hoover maintained secret files on all people in power in the government and elsewhere. He would uncover one or other piece of scandal and, without any threats or demands, let the victim know that he was in possession of the incriminating evidence. This would be enough to keep the victim from doing anything that could possibly compromise Hoover. It explains why he stayed on as the director of the FBI for so long, and it more than explains the famous comment by Lyndon B. Johnson that he 'would rather have Hoover inside his tent pissing out than outside his tent pissing in'.

For whatever reason, the FBI neither investigated organised crime in the early part of its history, nor recognised the existence of gangs. During Hoover's tenure the bureau never even admitted that there *was* organised crime, although it was the fastest-developing crime problem in the US at the time. If one is to believe such writers as Anthony Summers in his bestseller *Official and Confidential: The Secret Life of J. Edgar Hoover* and Curt Gentry in *J. Edgar Hoover: The Man and the Secrets*, the reason for this blind spot was that the Mafia, in the person of Meyer Lansky, had obtained compromising photographs of Hoover that showed his homosexuality. Hoover was, in effect, hoist by his own petard.

The point is that the FBI was dominated by Hoover; its policies were dictated from the top. Image was everything, to the extent that the entire operation of the organisation and its scientific policies were conducted in complete secrecy. Such behaviour is inimical to good scientific methods, which thrive on openness and debate.

This way of operating is found in all hierarchical structures, not least in South Africa, when the forensic laboratory of the old SAP fell under the authoritarian and dishonest Lothar Neethling. Described by one of the Supreme Court judges as reminding him of Josef Mengele, the notorious Nazi concentration-camp doctor who experimented on inmates, Neethling faithfully served his political masters to the detriment of forensic science in this country. He furthermore ran the state forensic laboratory in conditions of complete secrecy.

It is against this background of fraudulent fingerprint misidentifications and the workings of controlling, hierarchical organisations that I explore the case of Brandon Mayfield.

Chapter 6

THE MAYFIELD DEBACLE AND OTHER DISASTERS

'The scientist is not a person who gives the right answers, he's one who asks the right questions.'

– Claude Lévi-Strauss

'In questions of science, the authority of a thousand is not worth the humble reasoning of a single individual.'

– Galileo Galilei

'All men make mistakes, but only wise men learn from their mistakes.'

– Winston Churchill

Brandon Mayfield was a home-grown American boy born in Oregon and raised in Kansas. He served in the US Army and met and married an Egyptian national, whom he had met in Washington, D.C. After his marriage, he converted to Islam, and he also graduated as an attorney. There are a number of factors that, rightly or wrongly, resulted in heightened FBI suspicions of the man. Not least was the fact that he had converted to Islam. It was also recorded that he had offered legal assistance to a man called Jeffrey Battle, one of a group of men convicted of attempting to assist the Taliban. These unrelated issues all played a role in the subsequent fiasco.

On 11 March 2004, a bomb exploded on a Madrid train, resulting in many injuries and a final death toll of 191 people. The Spanish police obtained digital images of partial prints lifted from plastic bags found at the scene of the crime. The bags contained detonators. In due course, these digital images were sent to the FBI by the Spanish authorities. The FBI passed them through the multimillion-dollar Automated Fingerprint Identification System (AFIS) and, out of a database of almost fifty million prints, the submitted fingerprint was deemed to be similar to about fifteen sets. At this point, human fingerprint examiners compared the partial Madrid print to these fifteen possibilities. The FBI and the Spanish police agreed that there were eight points of correspondence between this partial print and that of Brandon Mayfield. However, the Spaniards were not convinced that the print belonged to Mayfield – there were also significant differences.

American paranoia was in full cry at this point in time, and it would appear that the federal authorities placed Mayfield and his family under visual and wiretap surveillance. It would also seem that there were intrusions into his house, with no items stolen, although this was never proved. Significantly, Mayfield had not left American shores for eleven years – something that seems to have bypassed the authorities. By this stage, the Spanish police had positively excluded Mayfield's prints. By 19 May 2004, they had provided the FBI legal attaché in Madrid with a letter confirming that they had identified the latent print as belonging to another person, namely Ouhnane Daoud, an Algerian who had become a suspect in the investigation for other reasons.

In the interim, the FBI had arrested Mayfield. The arrest was made in terms of a material-witness warrant rather than an indictment. Mayfield was held without access to adequate legal representation, and his family was informed neither of his whereabouts nor of the reason for his detention. Furthermore, he was held under a false name.

When the FBI finally and grudgingly realised that they had made a mistake, Mayfield was released. He sued the government successfully and was awarded damages of two million dollars.

The FBI, despite all the warning signals, were 100 per cent confident of the 'match', despite the Spanish misgivings and despite the fact that Mayfield had not left the continent, which meant that he could not possibly have been in Madrid at the time of the bombing. Their arrest and detainment of Mayfield was done in such a way as to deprive him of his constitutional rights and caused maximum distress to his family – all of this in the land of the free.

The problem started with an incompetent examiner not heeding the warning evidence. In addition, the number of similarities between the latent print and Mayfield's print was not sufficient to provide a match (in spite of this, the FBI examiner declared 'a definitive match' at one point). However, the team of FBI print examiners has much more to answer for: they must have known about the doubts of the Spanish authorities, but collegiate loyalty and confirmational bias trumped good, objective science. The subsequent examinations of Mayfield's latent prints were carried out by men who knew his history and background. They must also have been aware of the behaviour of their brother agents with respect to the surveillance and unconstitutional detention of the suspect. The probability that cognitive bias played a significant role in the ensuing debacle is high. Writing in the *Brooklyn Law Review*, Jennifer L. Mnookin puts the issue in the following way: 'Once the first FBI examiner declared the prints to match, the verifying examiners expected to find a match. It is no great surprise, then, that they found precisely what they expected to find, likely the result of a mixture of peer pressure and expectation bias.'

In 2006, a paper by cognitive psychologist Itiel E. Dror and his colleagues David Charlton and Ailsa E. Péron appeared in the up-market forensic science journal *Forensic Science International.* The paper was titled 'Contextual Information Renders Experts Vulnerable to Making Erroneous Identifications'. What Dror did was to take a group of five fingerprint examiners and give them a set of latent prints together with a suspect source print. The examiners were told that these were prints from the Mayfield case, which by now was well known and somewhat notorious. These hapless examiners were asked to establish whether or not there was a match. So far, so good. What they were not told was that, far from being a set of prints from Mayfield's case, these were actual prints and latents that they had previously examined in other, unrelated cases and which they had declared 'matches with 100% certainty'.

In Dror's examination, 60 per cent of the examiners reached the opposite conclusion, declaring no match. One of the five declared that the match was inconclusive, and only one came to the conclusion that he had previously reached. So much for 100 per cent accuracy claimed by fingerprint experts worldwide. Mnookin is quite right when she says that 'Cognitive biases are an inherent danger of our cognitive architecture' and 'that forensic scientists are not immune to them is hardly a surprise – except, perhaps, to those forensic scientists who were committed to a conception of their infallibility'.

This is frightening research for a number of reasons. Firstly, so little research has been carried out in an area of such immense importance in forensic science. Secondly, the establishment has reacted quite negatively to this sort of research. Thirdly, our legal system is ill-equipped to deal with the uncertainties inherent in science, preferring to create certainties and simplicity where neither exists. Forensic scientists and forensic pathologists come into the field and are not given the type of training that would predispose them to ask searching questions. Indeed, most of those who are in the state or prosecution ranks are committed to the idea that they are infallible.

The point is made even more convincingly by the response from the FBI in May 2004 once the scale and extent of the erroneous identification in the Mayfield case became known. I have included the FBI's press release on the Mayfield debacle in its entirety:

> After the March terrorist attacks on commuter trains in Madrid, digital images of partial latent fingerprints obtained from plastic bags that contained detonator caps were submitted by Spanish authorities to the FBI for analysis. The submitted images were searched through the Integrated Automated Fingerprint Identification System (IAFIS). An IAFIS search compares an unknown print to a database of millions of known prints. The result of an IAFIS search produces a short list of potential matches. A trained fingerprint examiner then takes the short list of possible matches and performs an examination to determine whether the unknown print matches a known print in the database.
>
> Using standard protocols and methodologies, FBI fingerprint examiners determined that the latent fingerprint was of value for identification purposes. This print was subsequently linked to Brandon Mayfield. That association was independently analyzed and the results were confirmed by an outside experienced fingerprint expert.
>
> Soon after the submitted fingerprint was associated with Mr. Mayfield, Spanish authorities alerted the FBI to additional information that cast doubt on the findings. As a result, the FBI sent two fingerprint examiners to Madrid, who compared the image the FBI had been provided to the image the Spanish authorities had. Upon review it was determined that the FBI identification was based on an image of substandard quality, which was particularly problematic because of the remarkable number of points of similarity between Mr. Mayfield's prints and the print details in the images submitted to the FBI.
>
> The FBI's Latent Fingerprint Unit has reviewed its practices and adopted new guidelines for all examiners receiving latent print images when the original evidence is not included.

A more abjectly dishonest piece of spin-doctoring would be difficult to imagine, and that from the so-called top experts in the field.

In an unusual move, the FBI conducted its own investigation into the affair. It appointed a panel of independent experts who concluded that 'the problem lay not in the quality of the digital images reviewed, but rather in the bias and "circular reasoning" of the FBI examiners'. Rather different from their press release, I would say.

Some idea of the uphill battle experienced by those of us who would like to inject a little more science into forensic science may be gleaned from the comments made by contributors to an online chat site. The topic under discussion was 'Why Fingerprints Aren't the Proof We Thought They Were'. John O. posted: 'This is an example of the left thinking college professors or forensic experts who never examined a latent print.' W.W. replied: 'There are no two fingerprints from different fingers alike and never will be. I have been doing this work for over 45 years, and no, I've never made a mistake in a fingerprint identification. NEVER.'

Quite why the first writer sees this as a left-wing issue is beyond me, and clearly informs us more about the writer than it does about the subject. The second contributor, who is all too typical of the genre, is one of those who knows not and knows not that he knows not. He is a fool to post this comment. Unfortunately, there are many like him.

As for the prints themselves, the similarities between Ouhnane Daoud's print and that of Mayfield are striking, but the differences are there. When one has view of the actual print left in Madrid, one can understand the difficulties. Yet the differences were not spotted by the group that boasts to be the world's best and which claims – or claimed – 100 per cent precision.

Ultimately these issues play out in a courtroom setting, and the matter is only of relevance to the real world if the courts can understand the difficulties and act on that understanding to give life to the problem. Despite numerous attempts on the part of attorneys to raise these issues, only one court has grasped the nettle. This was in 2007 when a court in Maryland excluded fingerprint evidence, calling it a 'subjective, untested, unverifiable identification procedure which purports to be infallible'. In this matter the circuit judge, Susan M. Souder, got it right. Unfortunately, the decision did not stand for long. In 2009, it was overturned by a federal judge, who said, 'Fingerprint identification evidence based on the ACE-V methodology is generally accepted in the relevant scientific community, has a very low incidence of erroneous misidentification and is sufficiently reliable to be advisable.'

What the learned judge failed to grasp is that the 'scientific community' that finds this method generally acceptable has a very low level of qualified scientists, as members consist mainly of police personnel. The judge also failed to understand that this so-called scientific community has a vested interest in supporting the notion of fingerprints being an unassailable identification method. The most glaring error, however, is that the incidence of misidentification and error is not known, so the judge could never have been able to justify his notion of a 'very low incidence of erroneous misidentification'. The 'methodology' of ACE-V is dealt with in more detail in Chapter 17.

Debacles in the world of friction-ridge analysis are not confined to the high-profile international cases dealt with above. They played a major role in the matter of *State* v. *Van der Vyver*, mentioned briefly in Chapter 3, where the misinterpretations of prints almost sent a young man to prison for life for a crime he did not commit.

On 16 March 2005, Fred van der Vyver said goodbye to his girlfriend, Inge Lotz. They were both students of mathematics at Stellenbosch University. Fred left Inge's townhouse, where he had spent the night, and after purchasing an item of furniture for a friend, he set off in his light delivery van for Old Mutual, the financial investment group in Pinelands where he worked, about an hour's drive from Stellenbosch. There he clocked in by way of the video–card identification system, and he was subsequently in meetings until just after 17:00. At or around 18:00 he was at his computer work station, as was shown by the emails he sent and by the testimony of a co-worker at the next station on that floor. Fred then left the building some time after 18:15 and went to his flat nearby. He delivered the item of furniture to his friend, had supper and made a few phone calls. From about 20:00 that evening he started to become concerned about his inability to raise Inge on the phone. This culminated, at around 22:00, in his sending someone to Inge's home to investigate. The news was not good. Between 22:45 and 23:00, Inge's lifeless body was discovered in her flat on her couch in the lounge. She had been brutally bludgeoned to death, as well as stabbed repeatedly. This melancholy tale ended in Fred's arrest and in his standing trial for murder.

I do not intend to deal with all the evidence against Fred here; that has been dealt with in my book *Steeped in Blood* and in *Fruit of a Poisoned Tree* by Antony Altbeker. The evidence at the trial shows first how prescient and perspicacious Nietzsche was when he wrote: 'Convictions are more dangerous enemies of the truth than lies.' Suffice it to say that all the evidence led by the prosecutors turned out to have been fatally tainted by a combination of incompetence, stupidity and fraud. Right now, I would like to deal with the fingerprint evidence only.

After saying goodbye to Fred, Inge's movements are not clear until about lunchtime, when she met up with a long-standing friend named Wimpie Boshoff. After lunch, we lose sight of her until about 15:00, when she bought a hamburger and a cold drink from a local fast-food outlet about three or four kilometres away from her townhouse. At around the same time, she hired a DVD from her DVD store, in the same shopping complex. We know that she returned home and was seen going up to her flat just after 16:00. That is the last time Inge Lotz was seen alive. The next event was, of course, the discovery of her mutilated body. Preliminary investigations revealed that nothing of value was missing from the premises. It was also noted that there was no forced entry and that the front door was open at the time the body was discovered.

Inge was found on the couch in the sitting room with a magazine lying on her lap. She was wearing a light T-shirt and flimsy shorts, the type often used as sleepwear. Several inferences could be drawn early on. Firstly, the person with her in the flat was known to her. Secondly, Inge was comfortable in his or her presence clad in such skimpy attire. Thirdly, the vehemence of the attack and the repeated injuries to the girl implied a violent act of great emotional anguish and passion. Suspicion, as we know, fell on Fred.

It needs to be said that, apart from circumstantial evidence – such as the killer being known to and on intimate terms with the deceased, as well as some evidence of an emotional upset between Fred and Inge – the only forensic evidence that ever stood any chance of securing a conviction was so shot through with incompetence and dishonesty that the prosecutors should never have continued with the prosecution.

A fingerprint was allegedly found on the cover of the DVD that Inge had rented at about 15:00. The state alleged that, for Fred's fingerprint to be on the DVD, he had to have been in Inge's apartment after 15:00 – something that Fred denied.

Everybody who looked at the print was in agreement that it was Fred's. If it could be shown that it was indeed Fred's print on the DVD, then his defence would have been a non-starter. Questions would have been asked about why he was at Inge's flat and, if so, why he lied about it.

When I first saw the fingerprint and photographs of it (see photograph section), I was accompanied by a former South African policeman named Nico Kotze. We were both concerned about one particular feature on the lifted print. On the top and at the bottom of it, two obvious curved lines were visible. They were spaced about eighty millimetres apart.

At this point, a little more information about fingerprints and fingerprint lifts is essential. As described in Chapter 4, when an individual touches a surface with a finger or a hand, the sweaty secretions on the surface of the skin are deposited onto the touched surface, where they take up the pattern of the skin. This deposit is the latent print, and these can often be seen on glass or shiny surfaces. To lift the fingerprint is a skilled and painstaking process.

The print is lightly dusted with a very soft brush (not unlike an artist's paint brush). The brush conveys minute particles of aluminium dust that stick preferentially to the oily substances in the latent print. But the aluminium powder also sticks to the surface on which the fingerprint has been deposited (the substrate), so when the fingerprint is lifted, some of the fingerprint powder on the substrate is lifted as well. This means that there will be some background around the lifted print.

The fingerprint expert, having dusted the prints, has to convert the dusted prints into a permanent record. This he or she does by applying a broad piece of sticky material to the surface where the latent has been dusted. The fingerprint powder will stick to this sticky material, and the lifting tape can be preserved and photographed.

There are several methods for lifting the print. Broad Scotch tape works very well, and the authorities are fond of using a sticky gel-type substance called foline on a black cardboard base. Whichever method is used, the principles stay the same. First, it is important to photograph the dusted latent print, as this gives context to the print. Second, it is vital to record the date, time and exact position of the lift. Third, if possible, it is desirable to retain the item on which the print was found.

In the case of the print called Foline 1 in the Inge Lotz case, none of the above precautions were observed. The importance of this single print cannot be overstated: Fred's fingerprint on the DVD cover would almost certainly have sent him to prison for a very long time. So what did the police do – or rather, fail to do?

They did not photograph the DVD cover with the dusted print in place. They never marked the print on the scene, thus damaging its provenance irreparably. And, arguably the most stupidly, they gave the

DVD cover back to the shop, destroying for all time the possibility of establishing the print's true origin.

The fact that the police returned the DVD cover could have been a crass act of stupidity, but I think this is unlikely. When the police confronted Fred during the interrogation, he was aware of the allegation that his fingerprint was on the DVD cover. If he had at that stage demanded the DVD cover, it would have been child's play to establish whether or not the print on Foline 1 had come from the DVD. I believe that the police knew that the print was from elsewhere and, rather than being caught in a barefaced lie in such a high-profile case, they got rid of the evidence by returning the cover to the DVD store and deciding to busk their way through the storm if and when it broke.

The fact that there were parallel curved lines about eighty millimetres apart on the lift known as Foline 1 was concerning. The parallel lines reflect something on the substrate. The only place where any lines could have been found on the DVD cover was on the edges. These lines, however, were neither curved nor eighty millimetres apart, and they demonstrated a 'tramline' quality that was not present on the Foline 1 lift.

Very early on, when I saw the fingerprint, Kotze and I were alerted to this problem. We spent a fruitless afternoon in August 2005 in Inge's flat searching for the surface from which Foline 1 was lifted. About a month later, another fingerprint expert, Daan Bekker (now deceased), suggested that the print on Foline 1 had been lifted from a glass.

Had Constable Swartz, the young policeman tasked with lifting the fingerprints at the crime scene, followed standing operating procedures and marked up the fingerprint lifts, this may have reduced the confusion. As it turned out, the prints were only annotated days after they were lifted – another factor in reducing the provenance of the fingerprint.

The fingerprint evidence took up a substantial part of the trial. The prosecutor blew hot and cold over the print. In his opening address to the court, he informed the judge that he would not be relying on the fingerprint, yet he led copious evidence about it and ended up trying to use it to obtain a conviction. The ploy was nothing if not sailing very close to the wind as far as integrity and honesty are concerned. The fact that the prosecution knew about the failure of Constable Swartz to mark up his fingerprint lifts and yet did not inform the defence about this important point does nothing to dispel the view I have of prosecutorial practice. In the end, it required the combined efforts of Pat Wertheim, Arie Zeelenberg and Paul Ryder to undo the evidence of the fingerprint, which the prosecutor knew from the outset was never going to stand up in court. Had Fred not been able to muster considerable resources, he would today be sitting out his lengthy sentence for a crime he could not have committed.

The cases of Brandon Mayfield and Fred van der Vyver are not isolated as far as erroneous fingerprint identifications are concerned. In January 2004, Stephan Cowans was released from prison in Massachusetts after serving six and a half years of a forty-five-year sentence for shooting a police officer. He was convicted on eyewitness testimony and fingerprint evidence, but was subsequently exonerated as a result of post-conviction DNA testing. The Boston Police Department later admitted that the fingerprint evidence was erroneous.

In Hatfield, US, a forensic technician used fingerprint impressions to identify a corpse. The individual identified as the corpse turned out to be alive.

There are many other cases like these. Even if we consider the brief list of faulty, mistaken and false fingerprint identifications that I have outlined above, the message is clear: no fingerprint evidence can be accepted unless it has been thoroughly tested by the defence and the court. The say-so of the police fingerprint expert is simply not enough.

Chapter 7

THE USE AND ABUSE OF STATISTICS IN THE COURTROOM

'There are three kinds of lies: lies, damned lies and statistics.'

– Benjamin Disraeli

'Facts are stubborn things, but statistics are more pliable.'

– Mark Twain

Some years ago, I was involved in the trial of several members of the then banned African National Congress (ANC). They were on trial for the capital offence of treason. Judge Daniels, who presided over the matter, showed scant interest in the defence case, slouching back on the judicial chair, his feet on the desk.

One of the issues deliberated was whether certain correspondence had been written by the accused. Giving evidence for the state was a senior police officer by the name of Hannes Hattingh, who was second in command in the questioned-document section of the SAP

Forensic Laboratory. Hattingh solemnly went through the handwriting contained in the documents in question, comparing each letter of the alphabet to known examples of the handwriting of the accused. Based on similarities of letter construction, he opined that the chances that the handwriting in the questioned document could have been written by anyone but the accused were in the order of one in hundreds of millions. This figure was calculated by taking the probabilities of each letter form for each letter of the alphabet – for instance, he would say that a particular form of the letter 'A' occurred only once in 500 different types of handwriting, and so on – and multiplying these probabilities together.

The fallacy in this argument is that the construction of individual letter forms are to a greater or lesser degree hammered into the young student, who copies one cursive script into a copybook when learning to write. It means, therefore, that the construction of each letter is not a totally independent variable. This can be illustrated quite easily. The handwriting of someone who was schooled on the European continent, especially in Germany prior to World War II, has a distinctiveness that is quite notable. Equally, the handwriting of older members of society has the mark of early copybook learning about it.

Hattingh's testimony is an example of overstating the evidence. It shows the abuse of a statistical calculation to enhance the evidence beyond its proper station. This kind of abuse of statistics to give greater value to a piece of testimony than it deserves is very common. Another example of such abuse is known as the prosecutor's fallacy, which makes use of a set of rules derived by Reverend Thomas Bayes, referred to as 'Bayesian statistics'. *Interpreting Evidence* uses the following example to illustrate the problem:

H = This animal is a cow

E = This animal has four legs

Then the probability $P(^E/_H) = 1$. That is to say, the probability that this animal that is a cow has four legs is 1 or 100 per cent (barring accidents where the cow has lost a leg). However, if the conditions are transposed – i.e. $P(^E/_H)$ – we can no longer say that the animal with four legs is necessarily a cow. It might be a sheep or a dog or a llama.

This is an example of the transposed conditional, also known as the prosecutor's fallacy. By transposing the condition, it is possible to change the probabilities in a given case erroneously. (In all fairness, defence lawyers make the same mistake – known as the 'defence attorney's fallacy'.)

Consider the description provided to me by Professor Patrick Randolph-Quinney of the Wits School of Anatomical Sciences:

> The prosecutor's fallacy is a mistaken inference of statistical probability made in law, where the context in which the accused has been brought to court is falsely assumed to be completely irrelevant in how the evidence is presented against them; the fallacy overplays the prosecutor's confidence in the evidence against the accused, and provides a statistical measure of doubt which is prone to over-emphasise the incriminating nature of evidence against the defendant. If the defendant was selected from a large group because of the evidence under consideration, then this fact should be included in weighing how incriminating that evidence is. Not to do so is called a base-rate fallacy.
>
> The fallacy usually assumes that the prior probability that a piece of evidence would implicate a randomly chosen member of the population is equal to the probability that it would implicate the defendant. The term was originally coined by William C. Thompson and Edward Schumann in their 1987 article 'Interpretation of Statistical Evidence in Criminal Trials: The Prosecutor's Fallacy and the Defense Attorney's Fallacy'.
>
> For example: A woman is mugged in a Johannesburg shopping centre, and her purse stolen. Inside the purse was R10 000 in cash, as well as personal possessions and bank cards. She, and several

eyewitnesses, describe the mugger as white, male, tall (over two metres) and in their twenties, with red hair.

Later that day the SAPS [South African Police Service] arrest a man matching this description who attempted to buy a laptop valued at R9 999 – suspicions were aroused when he attempted to pay for the laptop in cash, and in the same shopping centre where the mugging had earlier taken place. The man was charged with robbery and placed on trial.

During the trial, in the absence of any forensic or other physical evidence, the prosecutor attempts to link the defendant to the crime by showing how unlikely it would be (i.e. a low probability) that anyone else could have committed the crime. The prosecution lawyer produces an expert witness who quotes statistics from demographic data on the city. In delivering their evidence, the expert suggests that for the city of Johannesburg, the probability of each of the following physical characteristics of the defendant is as follows:

Being white = 0.23 (23% chance)

Being male = 0.51 (51% chance)

Being over two metres tall = 0.045 (4.5% chance)

Being between 20 and 30 years old = 0.25 (25% chance)

Being red-headed = 0.037 (3.7% chance)

He makes it clear to the court that because these variables are all independent of each other, we can multiply the probabilities together in order to obtain the probability of one person having all these characteristics: So, 0.23 x 0.51 x 0.045 x 0.25 x 0.037 = 0.00005

He makes the assertion that the chance of any random individual sharing all these characteristics is incredibly small – 0.00005 or 0.005% – therefore the chance of him being innocent is also incredibly small (0.005%). He damns the defendant by stating that because Johannesburg has a population of three million people, this means that there is a 1 in 20 000 chance of randomly choosing the defendant from within the GENERAL population (calculated by multiplying the population of 3 000 000 by 0.00005, and dividing 3 000 000 by the result). As such, a 1 in 20 000 chance of randomly finding our suspect will seem like long-odds to most people – therefore on the balance of probability he must be the thief.

Fine, except that this interpretation is completely fallacious – for the following reasons.

1. The probability of innocence is not 1 in 20 000. For the population of Johannesburg as a whole, a total of 150 individuals actually share the same suite of individual characteristics as our defendant. This is calculated from the first part of the equation used above – 3 000 000 x 0.00005 = 150.
2. This means that within Johannesburg 149 OTHER individuals are also walking around, sharing the same characteristics as the defendant. So, instead of a 1 in 20 000 chance of innocence, there is, in fact, a 149 in 150 chance (0.99 or 99%) that he DID NOT commit the crime – or, to put it another way, a 1 in 150 chance of being guilty.

It is worth noting that even though he is only 1 of 150 possible suspects this does not mean that he is innocent. The guilty man is one of these 150 individuals – and so it could still turn out to be our defendant. The police would have to exclude the other 149 tall, white, red-headed young males in order to be sure of the guilt of the defendant being tried.

The sad case of Sally Clark, which took place in 1998 in Britain, is a classic example of the problems besetting medicine and science coinciding: celebrity-witness status, overstating the evidence and venturing into expert fields beyond the competence threshold of the witness.

The Clark case is inextricably intertwined with the story of Samuel Roy Meadow. Born in 1933, Meadow was destined to rise to great heights and, as precipitously, plumb great depths. It would not be unfair to say that fame and infamy were his in almost equal measure. One is reminded of Thomas Gray's 'Ode on a Distant Prospect of Eton College':

> Ambition this shall tempt to rise,
> Then whirl the wretch from high
> To bitter scorn a sacrifice
> And grinning infamy
> The stings of falsehood those shall try
> And hard unkindness alter'd eye,
> That mocks the tear it forced to flow,
> And keen remorse with blood defiled,
> And moody madness laughing wild
> Amid severest woe.

Meadow studied medicine at Oxford and practised as a general practitioner (GP). Early on in his career he seemed to be a devotee of Anna Freud, the daughter of Sigmund Freud. His claims of a close association with Ms Freud have been met with denial by the Anna Freud Centre in London. In an interview conducted by David Cohen in the *Evening Standard* of 23 January 2004, members of Meadow's family were quite candid. In response to a question about Meadow's comment that he had been 'brought up' by Anna Freud, his ex-wife said, 'Anna Freud used to give regular seminars to paediatricians and the like and Roy occasionally went along. He had a great admiration for Anna Freud. There was something a little bit Bloomsbury-set about them – like an elite gathering of the great thinkers in child health of the time. But I have no idea why he said he was "brought up by her". That is stretching it a bit.'

After practising as a GP, Meadow trained as a paediatrician and in 1980 became professor of paediatrics and child health care at the University of Leeds. It was during this time that he came across a clinical case that was to have a significant impact on his life: in due course, it would make him famous and infamous in equal measure. The paper he published on this case propelled him into the limelight.

The paper actually describes two cases where young children were referred to the paediatric nephrology clinic in Leeds. In both cases, after lengthy and difficult investigations it appeared that the parents had provided 'fictitious information about their child's symptoms, tampered with the urine specimens to produce false results and interfered with hospital observations'. The outcome of the entire melancholy episode was that the police were called in when it appeared that both mothers were administering large doses of table salt to their offspring. The second of the two children eventually succumbed to the huge doses of salt and died. Meadow wrote up the cases and labelled the condition 'Münchausen Syndrome by Proxy'.

The name 'Münchausen' derives from a colourful character called Karl Friedrich Hieronymus Freiherr von Münchhausen, who was born in 1720. His claim to fame was that he had travelled very widely and spun strange and untruthful stories about these travels. His tales were subsequently published under the title *The Surprising Adventures of Baron Münchhausen*.

In 1951, a British physician called Richard Asher published a case history concerning a man named Thomas Beeches, who had presented to Harrow Hospital with a suspected abdominal obstruction. Exploratory surgery had shown nothing abnormal to explain the obstruction. During his work-up, Beeches had spun all manner of fantastic yarns about suffering massive abdominal injuries as a result of being torpedoed. He also told further stories about being a prisoner of war in Singapore. None of these stories was true. When his abdomen was examined, it showed a mass of scars of various vintage from previous operations. Asher coined the name 'Münchausen Syndrome' to describe Beeches's condition.

It comes as no surprise that, when Meadow was confronted with the little patients with puzzling and distressing symptoms and discovered that, in each case, the mother of the child had lied, he called the condition 'Münchausen Syndrome by Proxy'. The name rapidly became popular and doctors began to see much more of the condition. They were undoubtedly encouraged by Meadow's extensive publications list in all the right journals and books. There is no doubt that there are parents who do actual harm to their children in some misguided need to gain sympathy and attention. But was the condition as widespread as Meadow was contending?

The next step of the saga began when Meadow, by then a medical celebrity with a knighthood, entered the field of Sudden Infant Death Syndrome (SIDS). When he became involved in this rare and distressing clinical condition, it would seem that his mind automatically turned to Münchausen Syndrome by Proxy: his views were coloured by his obsession with the syndrome that had made him famous. Not only was this to lead to untold misery for the parents involved in the debacle, but it was to result in a catastrophic fall from grace for the man who started it all.

SIDS is usually referred to as 'cot death'. Its cause is not known, but it usually involves putting a normal, happy, bouncy baby to bed and returning to find that the child has died during the night, for no rhyme or reason. The causes of this condition are poorly understood. Its incidence varies, but averages two to three per thousand live births. Social conditions may play a role, the evidence suggesting a rate as high as 5.9 per thousand in lower socio-economic groups. The list of theories on causation is extensive, but nothing is published under the heading 'Multiple Sudden Infant Deaths'. Forensic pathologist Sir Bernard Knight notes that 'statistically a mother will suffer a second cot death in her family every quarter of a million births but a number of families are reported where three or even four siblings have died inexplicably. There is always the possibility of foul play in such circumstances, *though some familial metabolic or other genetic disease – albeit obscure – is more likely*' (emphasis added). He specifically cautions against ascribing more sinister events to such deaths in the absence of clear evidence.

With this brief background on SIDS and Münchausen Syndrome by Proxy we can start to unravel the terrible tragedy that befell a young Cheshire couple, the Clarks. Their eleven-week-old son, Christopher, was found dead, an apparent victim of cot death. At autopsy, no cause of death could be demonstrated. There was absolutely no evidence of foul play or lack of parental care.

Putting the unhappy episode behind them, Sally Clark fell pregnant again and produced a bouncing second child called Harry. Joy turned to anguish when, in a repeat of the past, Harry was found dead at the age of eight weeks. This time around, however, there was no sympathy: suspicion was immediately cast on the Clarks. After an 'investigation', Sally Clark was arrested and charged with the murder of the two infants, despite a lack of physical forensic evidence to indicate anything other than a second cot death. Sally was eventually tried. She was found guilty and received a life sentence.

This came about almost exclusively as the result of evidence given for the prosecution by Meadow. Essentially he told the court that the occurrence of two cot deaths in one non-smoking, affluent family was so rare as to make it a virtual certainty that both babes had been murdered. Meadow declared that the chances of two cot deaths in such a family were about 1 in 73 000 000 or, put differently, one could expect a double cot death in England about once every century. The court was clearly star-struck by the evidence of Sir Roy. At the peak of his fame at the time, having described and published widely on Münchausen Syndrome by Proxy, he clearly had no motive to give false testimony.

There was just one problem. The statistic of 1 in 73 000 000 double cot deaths was not only wrong, it was so wrong that it now ranks as one of the most infamous misuses and misunderstandings of statistics in courtroom history.

Meadow had uncritically accepted a statistic gleaned from a survey done in England, the so-called 'Confidential Enquiry into Stillbirths and Deaths in Infancy' (CESDI), which was done over three years, from 1993 to 1996. The figure for single cot deaths was given as about 1 in 3 003 in the general population, but, in an affluent family where the mother was over twenty-six and neither parent smoked, the statistic increased significantly to 1 in 8 500.

What Meadow simplistically did was to take the statistic of 1 in 8 500 for a single cot death and multiply it by itself to get the erroneous figure of 1 in 73 000 000. This reasoning is incorrect statistically because it assumes that there is no link between the two events.

A detailed analysis of the CESDI reveals much more than the crude approach adopted by Meadow. It can be shown quite easily that if a first cot death has occurred in a family, the likelihood of a second cot death is somewhere between ten and twenty-two times greater than it is in the general population; in other words, the deaths are linked by some unknown physiological factor. Thus, the likelihood changes from 1 in 8 500 for the second cot death in the same family to about 1 in 425 – that is to say, about twenty times more likely than it would be for the population group that would have included the Clarkes. This reduces the extraordinary statistic of 1 in 73 000 000 to 1 in 3 825 000 (1 in 8 500 x 1 in 425) or, given the fact that there are about 650 000 live births in England per year, it would increase the number of double cot deaths in England to far more than the statistics Meadow was quoting. This is much more in keeping with the numbers logged by the Foundation for the Study of Infant Death, which notes approximately two cases of double cot death per year in Britain – a far cry from the 1 in 73 000 000 quoted so confidently by Meadow, and certainly more frequent than once per hundred years.

Meadow's incorrect statistics sent Sally Clark to prison for a crime she did not commit. There was no motive and no forensic evidence, just the simplistic statistic of a man who was statistically illiterate and hopelessly overconfident.

So where did it all go so wrong? Firstly, Meadow was not a statistician. He lacked the understanding of statistics to enable him to make any statistical pronouncement. Secondly, his celebrity status had blinded him to his own inadequacies in areas outside of his true expertise. Thirdly, the court, the jury and the prosecutor were overawed by his evidence. (Here we see shades of Bernard Spilsbury.) Fourthly, the Clarks, despite being lawyers themselves, never obtained the expert evidence required to place the statistics in their proper light. There were other worrying features too. During Sally Clark's trial, Meadow claimed to have found eighty-one cot deaths that he ascribed to murder. When asked for the data, he claimed to have destroyed the raw data and his notes on the cases ... a distinct impossibility, in my opinion. To have destroyed raw data concerning a field in which he had an overwhelming interest and was a leading researcher simply beggars belief. No honest scientific researcher would do that.

Something not generally known about Meadow is that, as a young man, he played a starring role in an amateur production of *The Crucible* by playwright Arthur Miller. In the strangest pre-emption of what was to come, Meadow played Judge Danforth, a character who is at the heart of the plot and is ultimately discredited in the play, in which he pursues mothers accused of witchcraft and the unnatural murder of children. In the play, Danforth becomes judge, jury and executioner.

Further disturbing features of Sally Clark's trial have emerged. The prosecution pathologist, Alan Williams, failed to disclose to the court or to the defence the fact that Christopher, Sally's first child, had been shown at autopsy to have been badly infected with *Staphylococcus aureus*, which had caused bacterial meningitis; more than enough to have significantly altered the evidence against her. Once the second cot death had occurred, Williams altered his first post-mortem report, removing bacterial meningitis as a cause of death.

One must remember that, at the time that Meadow testified at the Sally Clark trial, the eponymous 'Meadow's Law' was being ascribed to him. It is not certain whether he formulated the law, but he almost certainly used it as a rule of thumb: 'One cot death is tragic, two is suspicious and three is murder.' After the Sally Clark trial and widespread complaints from the Royal Statistical Society about Meadow's misuse and abuse of statistics in the case, the British courts altered their approach to expert evidence to some degree.

The mistake made by the witness for the prosecution, the jury and all the legal parties involved was to fall into the trap of false statistical inference by a man who was not qualified to testify about statistics. The statistics were based on false premises (the independence of two cot deaths in one family), but the actions of Alan Williams, both in failing to inform the court and the defence about the *Staphylococcus aureus* infection, and in his alteration of the conclusions in the first post-mortem report on Christopher, illustrate the worst of the problems with a prosecutorially minded forensic pathologist. We have seen this before.

Meadow and his foolish, ill-considered evidence was the cherry on the top of one of the worst miscarriages of justice in the UK in recent times. Sally Clark, however, was not the only victim of Meadow's ignorance. Others caught up in the falsities propounded and promulgated by the arrogant Meadow were Trupti Patel, a thirty-five-year-old pharmacist accused of murdering three of her babies, and Angela Cannings, who hailed from Wiltshire and was sent to prison for the murder of her three babies. Meadow played a prominent role in both of their convictions. Similarly, Donna Anthony was wrongfully accused and sentenced for the murder of her son and daughter. She was released after serving six long years.

If any good has come out of this, it is that Meadow stands today as totally discredited. Meadow's Law is seen for the piece of bogus science that it is. The law has now been changed in England so that no person can be convicted on the basis of a single piece of expert testimony alone. These changes may bring cold comfort to people like Patel, Cannings and Anthony, and they have brought no comfort to Sally Clark, who died eventually of a combination of depression and other contributing factors. She died a broken woman.

Chapter 8

THE WHOLE TRUTH AND NOTHING BUT THE TRUTH

'Who lies for you will lie against you.'

– Bosnian proverb

Anyone who has read a courtroom novel or seen a courtroom drama on television will be familiar with the words of the oath administered to a witness prior to the giving of evidence: 'I swear that my evidence shall be the truth, the whole truth and nothing but the truth so help me God.'

Having taken this oath, to lie is to commit the criminal offence of perjury. In our courts, as elsewhere, perjury carries a substantial punishment – as demonstrated in the trial of Shirley McKie (see Chapter 5). The other role players in a courtroom are also bound by this ethical rule – namely to adhere to the truth. The entire process is, ostensibly, to see to it that justice is done and is seen to be done. Unfortunately, this ideal is not always observed.

In the late 1980s, I was consulted in a matter where a young man was fatally wounded in Worcester in the Western Cape. There had been what was colloquially referred to by the police as *onrus* (unrest) in one of the black townships on the edge of the white town. During the unrest, a young man by the name of Ndzima was seen walking towards a contingent of policemen with his hands in the air. He posed no threat and was, in any event, unarmed. Without warning or provocation, one of the policemen raised his rifle and shot the young man in his chest. The matter resulted in an inquest and, unusually, for the time, the magistrate made a finding of murder.

The state attorney acting for the murderous policeman eventually made representations to the Director of Public Prosecutions and the criminal case simply disappeared. This left the civil case. The family of the deceased sued the Minister of Police for damages resulting from the unlawful killing of Ndzima. I was briefed to act as an expert witness by Lee Bozalek, who at that time was a senior member of staff at the Legal Resources Centre. He is now a judge at the Western Cape High Court. Advocate Joel Krige appeared for the family of the deceased.

The crucial issue at the trial was to determine if the deceased had been shot while approaching the police or if he had been shot in the back while escaping from arrest, as the police were alleging. The family asked Lionel Shelsey Smith, who had been head of the Department of Forensic Medicine at UCT and was at that stage retired and doing private work, to carry out a second post-mortem on the body. On 30 October 1985, he performed the second autopsy in Worcester in the presence of Dr T.J. van Heerden, who had done the first post-mortem examination just over two weeks before.

The most important findings of Smith's post-mortem report are summarised in paragraph 8 (a) ii:

> Wound left chest resembling an entrance wound, perpendicular, oval 3.0 cm by 2.0 cm with a 0.3 cm abrasion collar on anterior axillary line. This wound was continuous with the underlying projectile track inclined downwards medially and posteriorly, passing between fifth and sixth ribs, upper margin of sixth rib grooved, transfixing the upper and lower lobes of the left lung, passing through the tenth intercostal space at the back.

In paragraph 9, Smith notes that 'a few chips of what appeared to be cancellous bone approximately two millimetre square were found in the projectile tract in the lower lobe of left lung'.

Stripped of the medical jargon, the message is that the bullet damaged the rib on the way in and exited between the ribs on the way out. Bony fragments from the damaged rib in the front were found in the wound track. There is only one conclusion: this man was shot in the chest, from the front. The police version was clearly false.

When I met with Smith before the trial, he became aware of the policeman's name from the case papers. Suddenly, without any proper explanation, he did an about-turn on his report, effectively – and fatally – damaging the plaintiff's case. He said to me in chambers that he felt he could not 'drop' the policeman, as he knew him well and had socialised with him in former years. So much for honesty and fearlessly stating the facts in a court of law.

Needless to say, I lost all respect for Smith and never again consulted him on any cases in which I was involved. I wonder just how often he tailored his evidence to fit the needs of an interested party, which, in his case, was inevitably the police.

Over the years, I worked with a Cape Town pathologist by the name of Len Anstey, who ran into major problems with a fellow called Professor Nel, the head of forensic medicine at Tygerberg Hospital, just to the north of Cape Town. It is at the discretion of any pathologist conducting an autopsy to allow others to attend the post-mortem. Nel always refused Anstey's requests, which were made on behalf of the families of the victims being autopsied, forcing him to get a court order. Why he should make it as difficult as possible for the family to get representation is not surprising seeing that Nel seemed to take it upon himself to act for the police and prosecution in a most partial way. Although the problems that Nel gave Anstey were singular, he was not the only forensic pathologist affected by Nel's behaviour.

Nel loathed me. On several occasions he came to court to assist the prosecution to cross-examine me on the subject of alcohol production in blood taken from the accused in cases of DWI. What contribution Nel could make to the process was mystifying: his knowledge of biochemistry was rudimentary. He was, however, smart enough never to enter the witness box on the subject if there was any chance of my directing cross-examination at him.

Nel appeared as an expert in the inquest into the death of Ashley Kriel, mainly to assist in cross-examining me. As evident from the discussion in Chapter 3 of the inquest into the shooting, justice failed Ashley Kriel. The magistrate and the 'experts' who appeared for the police were wilfully and, in my view, excessively partisan, and that was allowed to pass as justice.

The dishonest collusion of key witnesses and legal players acting in the interests of the prosecution or state has resulted in the denial and obstruction of justice time and again. In the case of Ndzima, a policeman had wilfully murdered a man who was politically opposed to his and the government's views. The court and the prosecution conspired with the police to allow a false version of events to be recorded, and no one was punished for the crime.

Chapter 9

BEYOND REASONABLE DOUBT

'The role of the expert witness is not to provide the evidence which supports the case for the Crown nor for the defence, unless that opinion is objectively reached and has scientific validity.'

– *Forensic Pathology*, second edition

In criminal trials, the state prosecution has to show beyond reasonable doubt that the accused is guilty of the crime as charged. The accused has to show only that his version is possibly true, within reason. The nature of and issues surrounding the burden of proof are found in all elementary textbooks on evidence – *The South African Law of Evidence* and *Phipson on Evidence* are particularly useful.

There are several fundamental principles to the burden of proof, including the Latin phrase '*Ei qui affirmat non ei qui negat incumbit probatio*', which means that the burden of proof falls on he who alleges and not on he who denies. This is visible in the notion of the presumption of innocence, the so-called golden thread of the English criminal justice system and many others.

In recent years, with the increase in international 'terrorism', many countries have circumvented this ancient legal principle and descended into much more barbaric customs. Under Tony Blair,

Britain introduced detention without charge (something more associated with B.J. Vorster at the height of apartheid than the UK). The US has maintained Guantanamo Bay, where the notion of no detention without trial is long forgotten and evidence is regularly and routinely extracted by torture and the third degree. The onus of proof has been abused to such an extent that one can only observe its devolution with sadness and a degree of cynicism. I would like to deal with some of these cases, and I hope to demonstrate how that high-sounding Latin phrase has been turned on its head.

On 22 May 1987, Mr Justice T.R. Morling returned his letters patent to the Governor-General and Commander-in-Chief in Canberra. Morling had chaired a royal commission to look into the criminal conviction of Alice Lynne 'Lindy' Chamberlain, who had been convicted of the murder of her nine-week-old daughter, Azaria, at Ayers Rock (now Uluru) on 17 August 1980.

When the child disappeared during a family camping trip, Lindy and Michael Chamberlain fell under the suspicion of the Australian police. At the first coroner's inquest on 20 February 1981, Magistrate Denis Barritt returned a finding that the baby had been snatched by a dingo. The Northern Territory Police were not at all happy with this finding, and their investigation proceeded with added zeal. On 18 November 1981, an order was made in the Supreme Court that quashed the findings of the Barritt inquest and directed that another inquest be held. This was duly complied with and, in February 1982, the coroner in the second inquest, Mr G. Galvin, returned a verdict that saw Lindy charged with murder and her husband, Michael, charged with being an accessory to murder.

After a lengthy trial, both Chamberlains were found guilty. Lindy was sent to prison while Michael effectively received a suspended sentence. All appeals against these findings were turned down, and that is where the judicial process rested for four years. In 1986, a chance finding of a piece of Azaria's clothing at Ayers Rock verified Lindy's testimony and resulted in her immediate release from prison.

On 15 September 1988, the Court of Criminal Appeal overturned all convictions against Michael and Lindy Chamberlain.

There is no doubt that this case received inflamed media attention and, to some degree, public opinion was set against the Chamberlains from the start, not least by Michael's position as a pastor in the church of the Seventh-day Adventists.

The finding of Azaria's missing clothing was one of those pure chance happenings. In 1986, an English tourist by the name of David Brett fell to his death at Ayers Rock. It was some eight days before his body was recovered. It just so happened that the body had fallen into an area that was difficult to reach and was teeming with dingo lairs. Because some predation had taken place on Brett's body, the police went searching for missing bones. It was during this search that they found a small item of clothing, which was quickly recognised as Azaria's missing matinee jacket.

Things had obviously gone wrong in the trial. Apart from the usual lurid stories dreamt up by the press and the introduction of a strange bias relating to the Seventh-day Adventist Church, the most significant piece of evidence against the Chamberlains was the 'blood spatter' under the dashboard of the Holden Torana motor car owned and driven by the couple.

Much of the prosecution case against Lindy rested on the 'presence of blood' on the dashboard on the passenger side of the car. Quite specific claims were made about this so-called blood spatter. Firstly, it was alleged that the substance was blood. Secondly, the 'blood' was said to contain foetal haemoglobin.

Haemoglobin is the red pigment in blood. It resides in the red cells and is the principle molecule that the body uses for conveying oxygen from the lungs to different parts of the body. The oxygen physiology of the baby before it is born is different from its physiology after birth, when the infant starts to breathe independently. In utero, the foetus's special oxygen-carrying needs are accomplished by foetal haemoglobin, which is chemically quite distinct from adult haemoglobin. By about six months, the foetal haemoglobin has all been replaced with the adult form of the molecule. The substance found on the Chamberlains' car was therefore alleged to belong to a baby not much older than six months.

The third damaging allegation about the spatter under the dashboard was that it was arterial blood. In all animals with a circulatory system, the blood is effectively divided into two systems separated by capillaries. The blood pumped out of the heart is under high pressure. By contrast, the blood in the venous system, which returns blood to the heart, flows at a much lower pressure. A cut into a vein therefore produces a non-pulsatile, low-pressure flow of blood, while a cut into an artery produces a high-pressure spray. When a high-pressure arterial spray hits any surface, the pattern can usually be distinguished from the low-pressure dripping of blood from a vein.

On the basis of this evidence, Lindy was accused of cutting Azaria's throat with a pair of scissors. The severing of a carotid artery (the main artery, one of a pair supplying blood to the brain) was said to have produced the arterial high-pressure pattern found in the car.

These are seriously damaging allegations and were backed up by 'scientific proof'. The Chamberlains had limited access to resources and scientific expertise to refute this evidence at the time. The state experts who performed the tests were Joy Kuhl, a forensic biologist in the employ of the government, and Dr Simon Baxter, who was at the time a senior forensic biologist with the Health Commission of New South Wales and Kuhl's supervisor.

The tests performed by Kuhl and Baxter involved reacting the material found in the car – 'presumptive blood' – with a commercial antibody. It would be helpful to understand how this test works in practice. Antibodies are protein molecules produced by the immune system of an animal, usually in response to an infection. When the animal's immune system encounters something 'foreign', such as a bacterium or a virus, it has a particular response to that substance. One of the components of the response is to produce specific antibodies that can recognise the foreign substance and bind with it. The body's reaction is utilised to produce special antibodies that can be used in the laboratory to identify other proteins.

Foetal haemoglobin is injected into rabbits kept specifically for this purpose. The rabbits, on being confronted with injections of foetal haemoglobin, proceed to make antibodies against it. It is these antibodies that are harvested and used in identifying foetal haemoglobin in forensic and clinical settings. In both settings it is very important to be absolutely sure of the specificity of the antibody. If the antibody is non-specific (in other words, if it gives a positive reaction with foetal haemoglobin as well as a number of other specific positive reactions), any tests done with it would be invalidated. I have included the prosecution's address to the jury on this vital issue (see Appendix A), taken from Morling's report.

I have in my possession a copy of the letter from Dr Störiko and Dr Bandner, both from Behringwerke Marburg, to Stuart Tipple, the Chamberlains' solicitor, which describes the nature of the antiserum used in the case. Page 2 of the letter, dated 21 July 1983, reads: 'Behringwerke does not guarantee that the anti-haemoglobin F [haemoglobin F is foetal haemoglobin] antiserum will react only with haemoglobin F in all test conditions' and 'the antiserum against haemoglobin F of Behringwerke, therefore, is not suitable on its own for the identification of foetal/infant blood and adult blood' (see Appendix B). This alone would have rendered Kuhl's work meaningless.

As the Morling report makes clear, there were additional problems. Kuhl made use of an antiserum not commonly used in forensic science laboratories, and she carried out no testing on the serum prior to using it in this ultra-high-profile case. Under cross-examination during the subsequent Morling inquiry, Kuhl was forced to concede that, because of the cross-reactivity of the antiserum, approximately a third of her tests would have been worthless. These doubts were never mentioned at the trial. To make matters worse, Kuhl's tests were missing important controls, and her record-keeping was deficient in important aspects. In addition, it was standard practice at the time to destroy the plates on which the tests were conducted and not to keep any photographic evidence, thus depriving the defence of any opportunity to verify the results. To say that this work was carried out in an unprofessional, incompetent and sloppy manner is an understatement. The fact that Kuhl concealed certain doubts from the court and the defence is a travesty bordering on perjury, if not actual perjury.

The spray pattern under the dashboard was re-examined at Morling's request. After an extensive investigation, it appeared that the pattern had been caused by a sound-deadening bituminous compound that had been sprayed into the wheel well of the car at the time of manufacture. The spray had passed through a drain hole into the car and onto the underside of the dashboard.

In 1981, forensic pathologist Dr Anthony Jones examined the spray pattern under what he described as a poor stereo microscope and opined that the 'droplets were not absolutely characteristic of blood'. Nobody raised this at the trial – the Chamberlains simply did not have the resources to challenge the spray-pattern allegations, a common theme in criminal trials. Jones did not see the signs of paint over the drops until much later, when he examined them under a decent microscope at the Victoria Police Forensic Science Laboratory. Why he did not have access to a good-quality microscope initially is mystifying.

During Kuhl's original testing of the scissors, several of the control samples failed to react. Four experts who appeared during the Morling inquiry for the defence were adamant that, under these circumstances, all the test results should have been 'reflected as worthless'. When confronted with this, Kuhl attempted to explain away the problem with some pseudo-scientific story. Ultimately, however, Baxter conceded that 'considering all the failures in the test, it should have been forgotten'.

Where were the checks and balances? Why did Baxter, her supervisor, not pick up the deficiencies before Kuhl gave evidence? Sloppy and unprofessional in the extreme is the only description that fits the bill.

Professor James Cameron, a pathologist and professor of forensic medicine at the London Hospital Medical College, claimed to be able to see 'impressions of the blood-stained hands of a small adult'. According to Morling's report, not only could no one else see these stains, including members of the High Court when the appeal was heard, but Cameron testified on the basis of photographs and some 'testing' done by Dr Andrew Scott (for the prosecution). Subsequent investigations showed that the 'blood-stained mark' seen by Cameron was, in fact, sand. Morling summarised the matter gently, in my view: 'It was unsatisfactory that his opinion was placed before the jury with the weight of his great experience behind it, without adequate verification of the results on which it was based.' I should say so. For Cameron to voice what must have been strong views given that this was a criminal trial can only be described in the harshest of terms, as the evidence was (for Cameron) second hand; he had not verified it himself. Above all, he could see what nobody else could. His willingness to abandon his role as an impartial, disinterested scientific witness is a permanent blemish on his name and reputation.

There were many other major travesties in the Crown's forensic evidence. In September 1980, Sergeant Cocks removed four hairs from Azaria's jumpsuit. These were handed to Dr Harding, a forensic biologist at the Adelaide Police Forensic Science Laboratory and witness for the prosecution. He examined these hairs and opined that they were probably cat hairs, but he would not deny the possibility that they were dog hairs. (Dog and dingo hairs are indistinguishable.) After the trial these hairs were examined by Hans Brunner, a senior technical officer at the Keith Turnbull Research Institute of the Department of Conservation and Lands in Frankston, Victoria, whose findings are contained in a report dated 18 September 1981. Again, Morling puts it in perspective: 'After the trial, the hairs were examined for the first time by Hans Brunner, an expert in animal hairs. He told the commission that, after using scientific methods unknown to Dr Harding, he had been able to demonstrate that the hairs were dog hairs. Dr Harding conceded that Mr Brunner was correct in his opinion.'

I will not dwell on the experiments that were carried out using the carcass of a goat that had been dressed in a jumpsuit and a nappy. To call them unscientific is possibly the kindest thing to be said about it all.

Where does this leave us? The most high-profile case in the history of Australia was pursued with extreme prosecutorial zeal. The state forensic experts were incompetent and, in the case of Kuhl, less than honest. The matter was marked by forensic experts who were quite prepared to act far outside their fields of expertise. This includes Dr Jones, who did not even insist on adequate equipment and who failed to see the overlying paint spots indicating that the so-called blood must have pre-dated the paint. Any self-respecting expert should have demanded and used proper equipment and, if that was not available, he or she should have refused to give an opinion. As for Cameron, he is the kind of pathologist who gives a really bad name to forensic medicine with this sort of testimony. Unfortunately, we have already met some others of his ilk and, sadly, we will meet a few more before this book ends.

The prosecutor in this case does not emerge unsullied either. He invited the jury to favour the evidence of Kuhl and Baxter over that of Professor Barry Boettcher, then a professor of biology in the Department of Biological Sciences at the University of Newcastle. For the prosecutor to make use of a shabby, invalid legal sleight of hand to have Boettcher's evidence disbelieved and the sub-standard evidence of Kuhl and Baxter supported is a major contributor to what has become one of the greatest miscarriages of justice of all time.

Chapter 10

THE *CSI* EFFECT AND REALITY

'There is no such thing as forensic science: instead it is a collection of scientific techniques and principles that are begged and borrowed from real sciences such as chemistry, biology, physics, medicine and mathematics.'

– *Encyclopedia of Forensic Sciences*

Popular television series like *CSI* and *NCIS* lead viewers to believe that forensic science will instantaneously shed light on impenetrable mysteries and will point investigators unerringly towards solving the crime.

I am afraid that this is TV and not reality. Forensic identification science, which contributes significantly to the forensic case load, is undergoing something of a re-evaluation. So much so that in 2009 the National Academy of Sciences in the US published the results of much soul-searching in this regard. The report-back, *Strengthening Forensic Science in the United States: A Path Forward*, responds to increasing evidence that things are not all that they should be in the field of forensic identification science. In the summary of their findings, the authors say that, 'in some cases, substantive information and testimony based on faulty forensic science may have contributed to wrongful convictions of innocent people'. The cases discussed in the previous chapters clearly illustrate this sentiment.

I would like to take you through some of this 'faulty forensic science' to show the extent of the junk science that has been touted in our courts of law, with incalculable effects on the lives of many people. If you were to take a simple introductory textbook on forensic science, such as Max Houck and Jay Siegel's *Fundamentals of Forensic Science*, one would encounter a wide variety of subjects, including determining the post-mortem interval (time after or since death), hair analysis, explosions, handwriting comparisons and ballistics. You might, after reading such a book, come away with the impression that the subjects are well grounded in basic science and that, barring anomalies, they are all clear-cut and easy to interpret.

Unfortunately, this is largely incorrect. Not only are the findings of forensic science experts vulnerable to cognitive and contextual bias, as shown in the preceding chapters, but the fundamentals of these subjects are not always properly grounded in science. It has long been thought that the legal process would weed out junk science and that cross-examination would provide a high road to the truth. Alas, this is not so. As *Strengthening Forensic Science in the United States* submits, 'The adversarial process relating to the admission and exclusion of scientific evidence is not suited to the task of finding "Scientific Truth". The judicial system is encumbered by, amongst other things, judges and lawyers who generally lack the scientific expertise necessary to comprehend and evaluate forensic evidence in an informed manner.' The authors were clearly aware of another major hindrance in this enterprise when they wrote, 'The forensic science enterprise is hindered by its extreme disaggregation – marked by multiple types of practitioners with different levels of education and training and different professional cultures and standards for performance … and a reliance on apprentice-type training and a guild-like structure of disciplines, which work against the goal of a single forensic science profession.'

Let us now examine some of the types of forensic 'science' that have come and gone without leaving a trace. A brief look at the literature will show that the name of B.D. Gaudette looms large.

Hair analysis

A paper by B.D. Gaudette published in the *Journal of Forensic Sciences* in 1982 discusses the statistical likelihood of a match between two randomly selected hairs. The figure that Gaudette used was 1 in 4 500 for two hairs from randomly chosen individuals, the inference being that if your hair matched one found at the scene of a crime, there was a 1 in 4 500 chance that it belonged to someone else. But cracks in the theory of the use of hair to identify criminals started to show. In a paper by P.D. Barnett and R.R. Ogle in the same journal, Gaudette's data is said to be 'seriously flawed' and fails to 'justify the statement that hair evidence is good evidence'. The subject is dealt with at some length in *Forensic Science Handbook* in an article by Richard E. Bisbing. In the article, Bisbing is much more circumspect than Gaudette in discussing probabilities.

Strengthening Forensic Science in the United States states: 'No scientifically accepted statistics exist about the frequency with which particular characteristics of hair are distributed in the population. There appear to be no uniform standards on the number of features on which hairs must agree before an examiner may declare a "match".' Later, the report declares: 'The committee found no scientific support for the use of hair comparisons for individualization in the absence of nuclear DNA.'

I have been consulted in cases in South Africa where the state's 'hair expert' measured only two or three different features of the hair under consideration and yet came up with the same statistic as Gaudette, namely 1 in 4 500. (Gaudette used about ninety different features.) How the police expert could come up with the same statistic using so few distinguishing features is a straight failure to comprehend the nature of the scientific method and the application of rational statistical evaluation.

Right up into the middle of the 1990s, hair analysts were using hair in order to secure convictions. It was only with the advent of DNA that the unreliability of hair to identify criminals was finally exposed.

In 1986 in Coral Springs, Florida, Michael Rivera was convicted on the evidence of a county hair analyst who identified a hair that had been found as coming from the head of the victim. It was some years later, after Rivera had spent the better part of two decades on death row, that DNA testing of the hair showed that it had not come from the head of the victim as originally argued by the 'expert' and the prosecutor. Similarly, Charles Fain was convicted of raping and murdering a nine-year-old girl in Nampa, Idaho, in the early 1980s. The conviction was based partly on hair identification by Michael P. Malone, a hair 'expert' in the employ of the FBI. Again, after nearly twenty years on death row, Fain was excluded as the source of the hairs by DNA analysis and walked free.

There have been many such cases. The analyst in the Fain matter, having testified in over 5 000 cases, was finally unmasked as unscientific and dishonest when his colleagues alerted the authorities to grave problems in Malone's laboratory. In true FBI-style, the whistle-blower, Frederic Whitehurst, was excoriated in an investigation while the FBI declined to act against Malone.

Yet again, we see the familiar cohort of poor science, over-reaching and less-than-honest expert witnesses, the failure to recognise the limits of the evidence, and institutionalised protection of sub-standard witnesses, which results in punishment of the innocent and, often, exoneration of the guilty.

Comparative bullet-lead analysis

Firearms are often used in criminal activities. They come in a bewildering variety of shapes, calibres, designs and models, though broadly they may be divided into rifles, shotguns and handguns. Rifles are usually high-power weapons that fire bullets at high velocities. They are more difficult to conceal than handguns and thus are less often used in criminal activities – except possibly in South Africa, where the AK-47 assault rifle is sometimes known as the 'African credit card', a reference to the frequency of its use in bank heists and cash-in-transit robberies. Handguns are altogether much easier to conceal and are the weapon of choice in most firearm-related crimes, while shotguns, which do not often feature in run-of-the-mill crimes, are sometimes used in suicides and home-based, inter-family murders.

There is a significant variation in bullet design. Many bullets have a central core of lead encased in a thin copper covering called the jacket. These are known as 'full metal jacket' bullets. Other bullets are simply cast from lead, while still others have a portion of the lead core exposed. The purpose of this is to allow the bullet to expand on hitting the target. The greater surface area slows the bullet down on impact, which releases more energy within the target – with correspondingly more damage and wounding potential. Where a hunter requires a knock-down shot or a police official desires more stopping power, this design feature is most appropriate. In addition, the expanding type of bullet prevents overpenetration and reduces the chances of wounding someone standing behind the primary target.

Depending on the type of weapon used, the only evidence that may be available to investigators after a crime has been committed is a bullet, removed either from the victim or from the scene of the shooting. The standard way of identifying the bullet and the weapon that fired it is to compare the bullet or fragments of the bullet under a comparison microscope. The exact nature of comparison microscopy must wait for Chapter 15; what I wish to deal with in this chapter is the notion that bullets may be identified and linked to a particular crime by means of analysing the micro-contaminants in the head of the bullet or its core and comparing these micro-contaminants to bullets found in the possession of the accused.

Often the bullet evidence recovered from a crime scene is so damaged that there is not enough of it on which to perform comparison microscopy. Bullet-lead analysis works on the principle that, if the trace elements in the lead in a bullet from a box in your possession match those trace elements in the evidence bullet, then you can be tied to the crime scene. There are good analytical methods for measuring all the minor metallic compounds in a sample of lead. The equipment back in 1960, when these methods were developed, was expensive and bulky. The method used then was neutron activation and it required, among other things, access to a nuclear reactor. The size and expense of the equipment limited the number of laboratories that could perform the work. Effectively, the only institution worldwide that regularly performed this analysis was the FBI.

The neutron-activation method is based on the following premises:

1. the fragments of the bullet are an accurate representation of the whole bullet;
2. the molten source of the lead from which the bullet is made is compositionally uniform; and
3. the composition of each molten source is absolutely unique.

If any of the above three assumptions are not valid, then the entire method and any inferences to be drawn from such a bullet-lead comparison analysis will be completely invalid.

By now it should be apparent that there is a pattern in this sort of junk science: poor-quality inferences, non-disclosure of problems with the method and denigration of whistle-blowers. It is instructional to follow the slow steps that led to the exposure of this method of bullet identification as bogus science and the ultimate abandonment of the method by the FBI.

In July 1995, a coin dealer by the name of Robert Rose was found dead. He had been shot. Suspicion quickly fell upon Michael Behn, who had been discussing a business deal with Rose earlier on that day. One of the factors leading to Behn's conviction was the evidence of Charles Peters of the FBI. Peters had measured the microcontaminant elements in the bullet from the deceased and found these to be in concordance with the same elements found in the lead in bullets retrieved from Behn's house. When Peters was asked about the probabilities of a bullet match by chance, he said, 'I can't [give] any statistics, but I could spend a lifetime looking for that.'

While Behn was sitting in prison, his sister heard of William Tobin, who had just retired after a career spanning twenty-seven years at the FBI as the chief metallurgist. Tobin had for a long time harboured serious doubts about the science of bullet-lead analysis. Incidentally, he had also harboured serious doubts about the FBI's Michael Malone, the so-called hair expert who was subsequently discredited, and he knew Whitehurst, who had blown the whistle on Malone. In fact, it was Whitehurst who had referred Behn's sister to Tobin.

Tobin was aware through personal communication with other laboratories that the chemical analysis of bullet lead was regarded as unreliable. These were not Mickey Mouse laboratories: they included the National Laboratory Center of the Bureau of Alcohol, Tobacco and Firearms in Rockville, Maryland, as well as the Bundeskriminalamt, the important forensic laboratory in Germany. After studying the neutron-activation method carefully, Tobin decided that it was extremely unreliable.

When Tobin was contacted by Behn's sister, he did what should have been done in the first place. He assembled a team of disinterested (non-government) scientists and they investigated the neutron-activation method from top to bottom. The clamour directed at this method eventually resulted in a study by the National Research Council, which issued a report titled 'Weighing Bullet Lead Evidence'.

The report was not supportive of the theory and basis of the comparative analysis of bullet lead. The FBI, not unexpectedly, issued a counter-report claiming that the science was sound and in general supporting the evidence provided by their agents. The FBI said: 'Published research and validation studies have continued to demonstrate the usefulness of the measurement of trace elements within bullet lead. The "science" has continually withstood legal challenges in state, local and criminal courts.'

After a short while, on 1 September 2005, the FBI, in a dishonest, face-saving press statement, announced that 'after extensive study and consideration' it would no longer conduct the examination of bullet lead.

Flowing from this announcement are all sorts of questions. Why was the 'extensive study and consideration' not done before this junk science was unleashed on the courts? Why was consultation not carried out with the alcohol, tobacco and firearms scientists or, for that matter, with the forensic laboratory in Germany? Why did the FBI, with all its resources, not go to bullet manufacturers such as Remington, where they would have found out that they produce bullets with up to fifteen different calibres from the same source of lead? This in itself invalidates the method. The system failed to such an extent that it allowed junk science into the courtroom, to the immense prejudice of many defendants, who had neither the resources nor the money to challenge it. Anyone interested in the fuller story of bullet-lead analysis and its appalling aftermath would do well to read Chapter 15 of Jim Fisher's *Forensics Under Fire*.

It is interesting to note that, had the FBI invested even moderate time and effort, they could have established that manufacturers like Remington and Winchester start off with a molten source that may be as large as 125 tons. The bullets formed from this mass of lead may number in the millions. For instance, if standard 9mm Parabellum bullets are being cast, they weigh about eight grams each. Thus, about 15.5 million bullets can be made from a 125-ton batch of lead. These bullets may be kept in stock or may be sent to large wholesalers or, for that matter, to other cartridge manufacturers. For the FBI agents to tell the courts that they could place the evidence bullet in a given batch of bullets and, in some cases, the same box, is not supported by even the most cursory investigation. Yet this was standard practice in the largest forensic laboratory in the US. What is more, the courts allowed this rubbish to pass muster.

The following chapter continues the discussion of bad science, and takes the debate further.

Chapter 11

BOMBING OUT IN BIRMINGHAM

'Seeking what is true is not seeking what is desirable.'

– Albert Camus

Society lurches between the fear of rampant criminality and the fear of convicting the wrong man or woman. It is often said, 'Better ten guilty go free than one innocent person be convicted.' This sentiment is attributed to the eighteenth-century English jurist William Blackstone, and is found in his magisterial commentaries on the laws of England. The ratio is of far less importance than the concept itself: that the law must always err on the side of innocence. Sadly, this notable aim is often lost in the hurly-burly of court practice, prosecutors' ambitions, defence counsels' scientific illiteracy and the criminal courts' tendency to lean towards the state cases – not to mention the dishonesty of the police.

England is often thought of as the place where the notion of jurisprudence germinated and took root. The Magna Carta was forced onto King John by his rebellious barons in 1215 and was the first great founding document delineating the freedom of the individual against the arbitrary whims of the throne. In due course, it spawned other documents that limited the powers of the authorities and forced them into abiding by the law. We need to examine the realities of everyday 'justice', however, and the behaviour of the police within the system, as well as the role of certain prosecutors in the miscarriage of justice. All too often, the quality of prosecution science is sub-standard. The inferences drawn from certain data are often overstated and, in some cases, totally unjustified.

The quintessential cases illustrating the level to which these noble ideals of justice, refined over centuries, had fallen by the 1970s are the criminal prosecutions that followed a series of bombings in the UK.

On 21 November 1974, just after 20:00, bombs went off in two pubs in Birmingham, the Mulberry Bush and the Tavern in the Town. Altogether twenty-one people died and 163 were injured. These explosions occurred just a week after the Irish Republican Army (IRA) had threatened to bring its war to England.

Within a short time, six men, who were to become known as the 'Birmingham Six', had been detained on their way from Birmingham to Belfast to attend the funeral of James McDade, a provisional IRA member who was accidentally killed while allegedly planting a bomb in Coventry. The men detained were Johnny Walker, Gerry Hunter, Hugh Callaghan, Paddy Hill, Richard McIlkenny and Billy Power.

Having been detained, their hands were swabbed by Dr Frank Skuse, who worked at the forensic laboratory in Chorley, England. The test he used is known as the Griess test. Named after Peter Griess, who first described it in 1858, the test has been in use for some time. A spot, or presumptive, test, it reveals the presence of nitrates, which are commonly found in explosives. Its limitations are and were well known in the world of organic chemistry. Anyone using the test will quickly discover that false positives are common; that is to say, a substance giving a positive response may very well not be an explosive.

The way in which the test works is as follows: The hands are wiped with a clean swab soaked in an organic solvent such as ether or ethyl alcohol. The swab is then treated with sodium hydroxide, also known as caustic soda. This has the effect of cleaving the nitrite group in the sample, leaving the nitrate group free on the swab to react with the Griess reagent, which is then mixed with the nitrate group, giving an attractive red-pink colour. The concentration of the caustic soda used in the test is of crucial importance.

Skuse used the Griess test on the hands of the detained men to determine whether or not they had been in contact with or handled explosives. He reported the presence of nitrates on their hands. The men were subsequently tried and convicted, and all received lengthy prison sentences.

The bombings resulted in an anti-Irish hysteria and the enactment of the Prevention of Terrorism Act, which passed through all the hoops in the House of Commons within a day and sped unamended through the House of Lords in three months flat. One of the features of this Act was that it allowed the police to detain suspects for up to five days, and thus was a serious encroachment on habeas corpus (the ancient order securing the release of a person from unjustifiable detention in prison, effectively meaning 'bring forth the person' for release or to stand trial), which had been part of justice since time immemorial.

The story of the miscarriage of justice ensuing from the arrests in these bombings has been told often (see, for instance, *Miscarriages of Justice* by Bob Woffinden). Where did British justice go off the rails? For the purposes of this book, I shall concentrate on where the forensic science went wrong and how the law dealt with the appalling lack of understanding and dishonest behaviour of the Crown experts.

With regard to the Griess test, any scientist worth his salt and familiar with the available literature of the day would have known that the test is non-specific. As a young student, I had a copy of the classical work on the subject, *Spot Tests in Organic Analysis*. This book made it abundantly clear that these tests were capable of detecting families of organic compounds and were in no way specific.

In addition, any competent scientist using the Griess test would have quickly found that many nitrate-containing compounds will provide a positive response. A good example is nitro-cellulose, which is used for high-gloss printing (of playing cards, among other things), as well as in propellants for the military. This is of special significance in the case of the Birmingham Six, in that the men played cards on the train just before their arrest and testing. One is tempted to suggest that all of this was mere ignorance and incompetence on the part of Skuse. I think, however, that the evidence points in another direction.

It appeared in the appeal hearings of the Birmingham Six that Skuse had adjusted the notes made by his assistant to ensure that the times of his tests squared with those of the police. Furthermore, Skuse gave evidence that, over and above the Griess test, he had used more sensitive and specific tests on the samples. He claimed to have used gas chromatography coupled with mass spectrometry. Gas chromatography mass spectrometry (GCMS) is a technique that separates mixtures of substances into their individual compounds and enables the scientist or chemist to identify the compounds with a high degree of accuracy (this technique is discussed in more detail in Chapter 14). The method produces a permanent printed record. Skuse's laboratory notebook should have recorded such tests, but, when challenged, Skuse could produce no documentation or instrument printouts. In a case of such importance, this is beyond belief and suggests that the additional tests were not done.

While carrying out research for a television documentary about the case, the scientists commissioned by the programme contacted Skuse for the recipe for his 'modified' Griess reagent. When the scientists used it, their results were quite different from the results originally obtained by Skuse. When he was confronted with this at an appeal, Skuse explained that he had given the television scientists a book recipe and not his own modification. It should be noted that if Skuse had indeed modified the method, there should have been

That 'incomparable witness', Sir Bernard Spilsbury

Dr Crippen and Ethel Le Neve stand trial in London

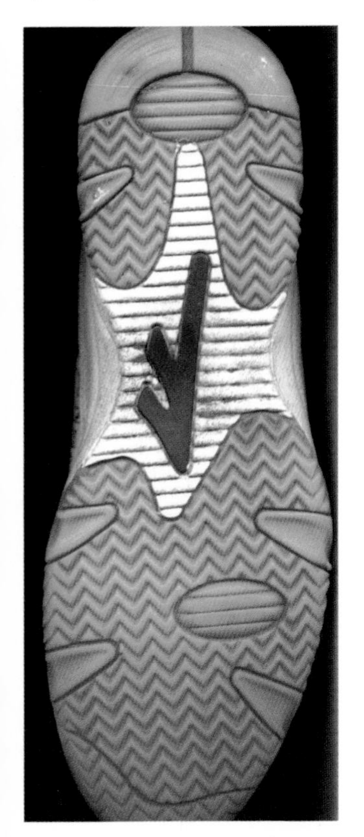

Fred van der Vyver's running shoe, which was alleged to have made the mark in the bathroom

The blood mark on the bathroom floor at Inge Lotz's townhouse

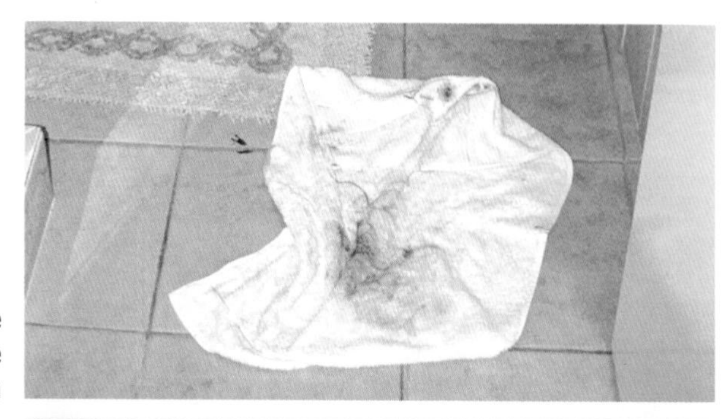

The bloodied towel and the so-called footprint in Inge Lotz's bathroom

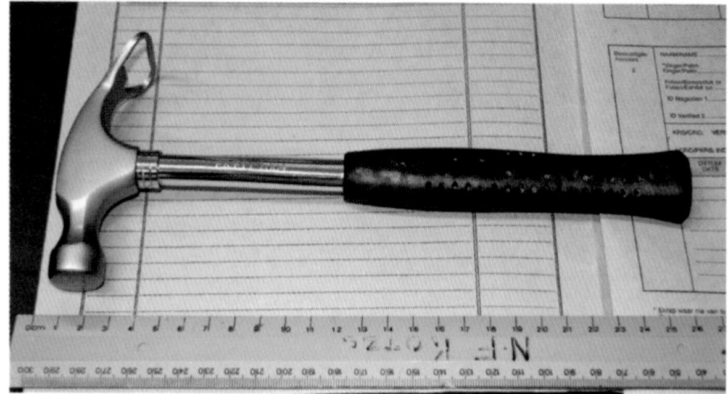

The alleged murder weapon in the Lotz case

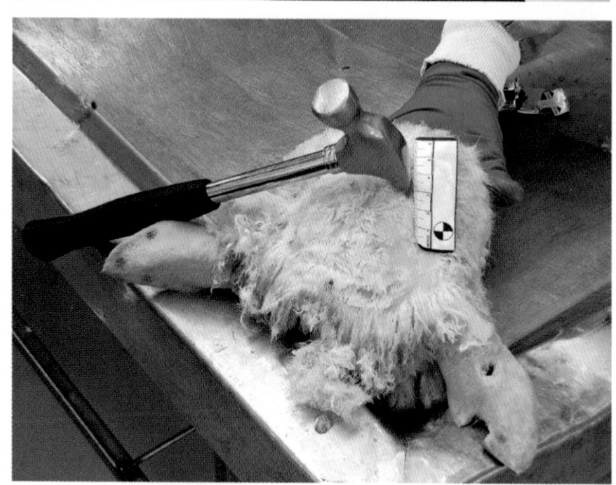

The police experiment to replicate the wounds to Inge's head

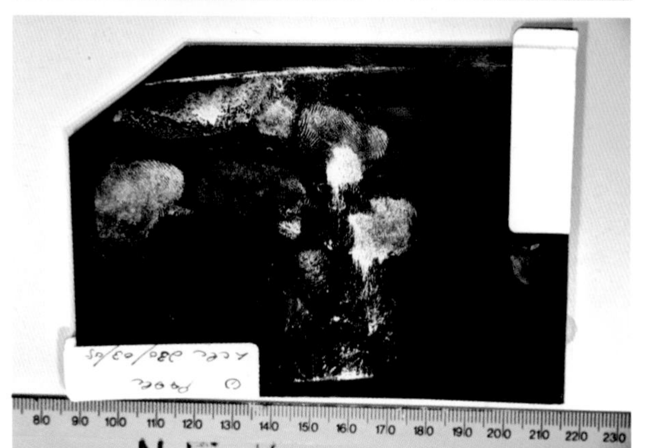

Fred's fingerprints – but were they lifted from a glass or a DVD case?

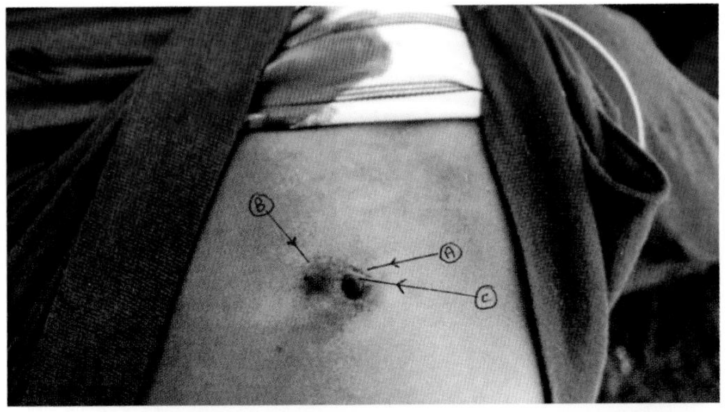

The bullet entrance wound to Ashley Kriel's body

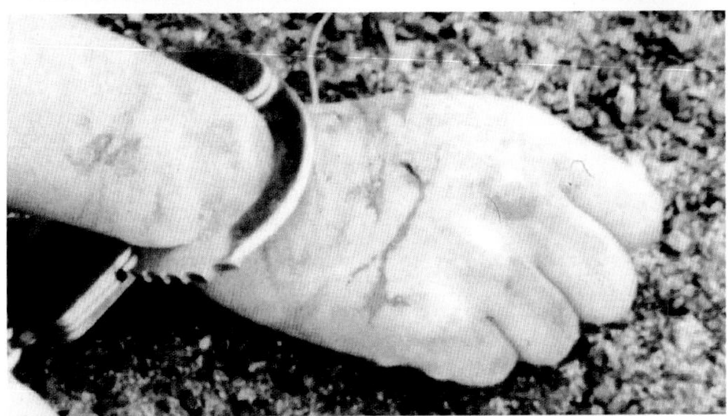

Kriel was handcuffed when he was shot

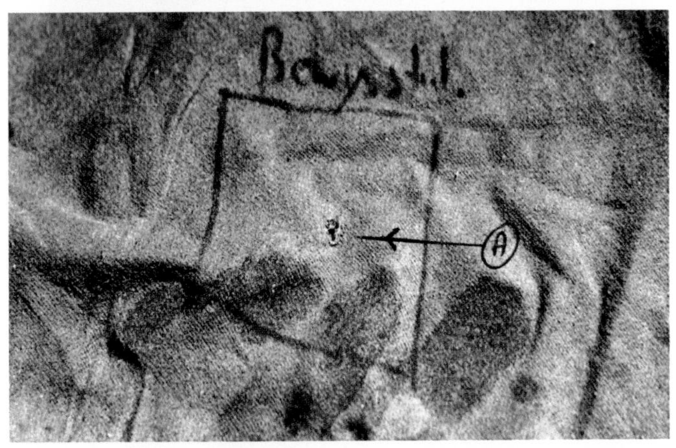

The bullet hole in Kriel's tracksuit

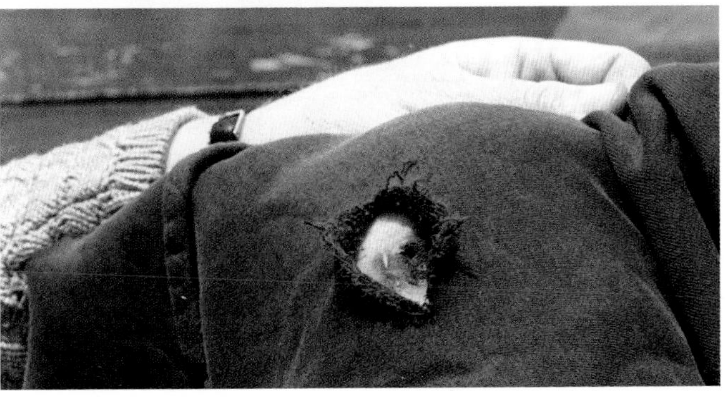

The damage caused to a tracksuit top by a contact shot

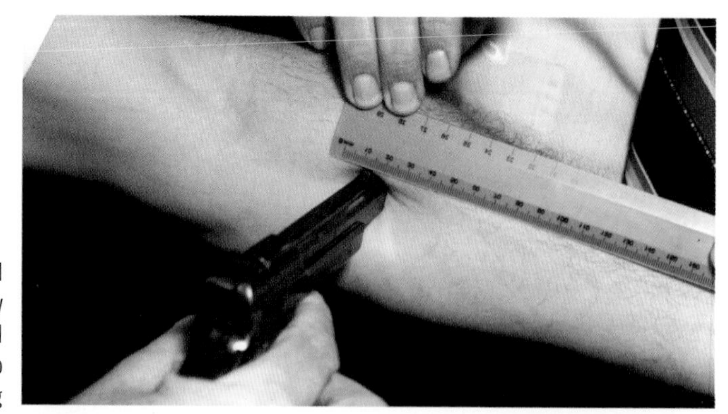

The nonsensical experiment performed by the state: remember, Kriel was shot through two layers of clothing

The weapon that killed Ashley Kriel

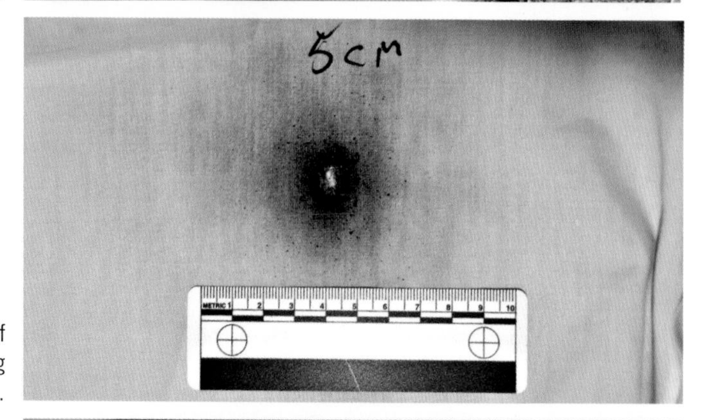

The typical appearance of a bullet hole in clothing fired from close range …

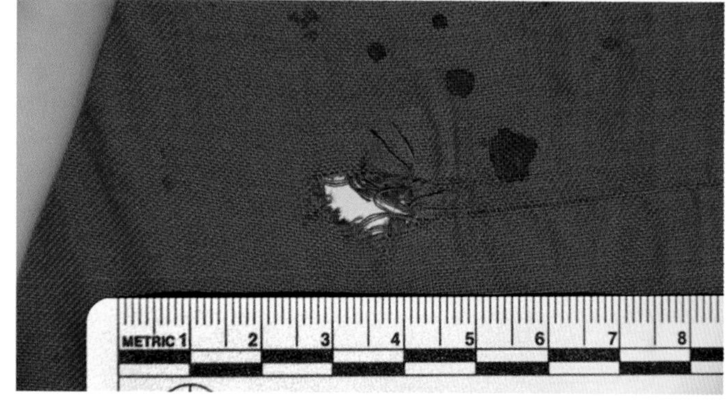

… compared to one the police produced, wrongly thinking the hole to have been caused by a bullet

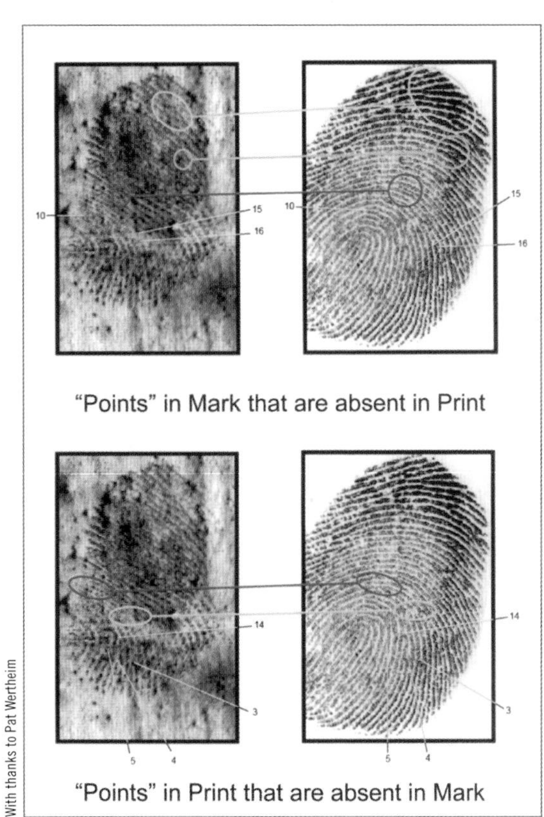

With thanks to Pat Wertheim

Fingerprint comparisons in the Shirley McKie case

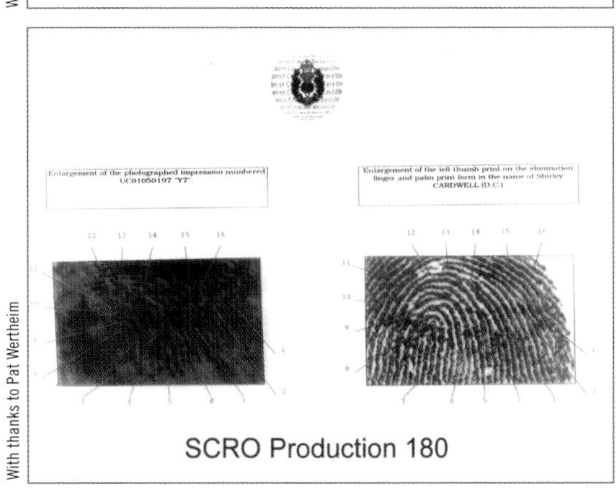

With thanks to Pat Wertheim

Shirley McKie's right-hand print (left), and on the right the print found at the crime scene

What your Lamborghini looks like after a fire

Ballistics comparison image 1: A good match from two bullets fired from the same gun. Compare this with the images that follow

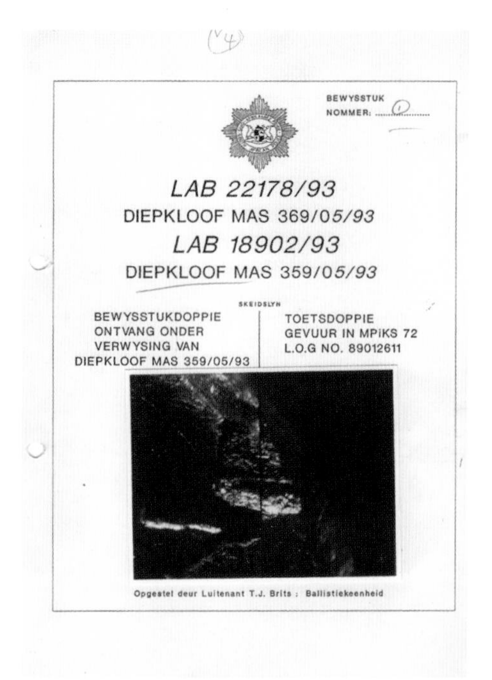

BEWYSSTUK NOMMER: 1

LAB 22178/93
DIEPKLOOF MAS 369/05/93
LAB 18902/93
DIEPKLOOF MAS 359/05/93

SKEIDSLYN

BEWYSSTUKDOPPIE ONTVANG ONDER VERWYSING VAN DIEPKLOOF MAS 359/05/93

TOETSDOPPIE GEVUUR IN MPIKS 72 L.O.G NO. 89012611

Opgestel deur Luitenant T.J. Brits : Ballistiekeenheid

Ballistics comparison image 2: A typical police 'match' with almost no identifying features

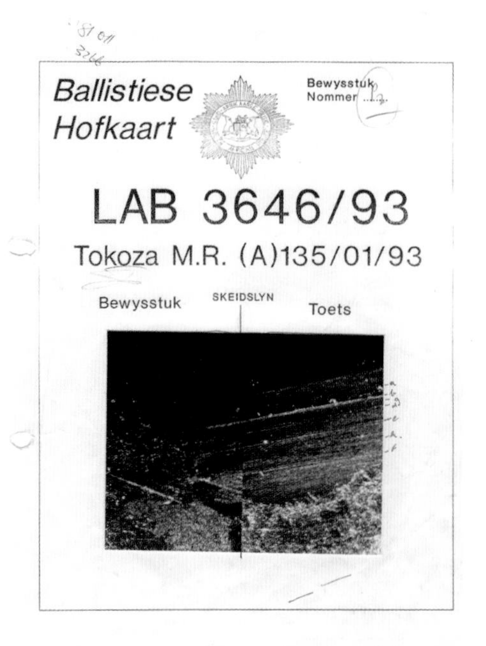

Ballistiese Hofkaart

Bewysstuk Nommer 1

LAB 3646/93
Tokoza M.R. (A)135/01/93

Bewysstuk | SKEIDSLYN | Toets

Ballistics comparison image 3: A typical police 'match'

Ballistics comparison image 4: This comparison reveals that one set of marks was made by a land (raised portion) in the barrel, and the other by a groove

Flintlock being discharged (top and above)

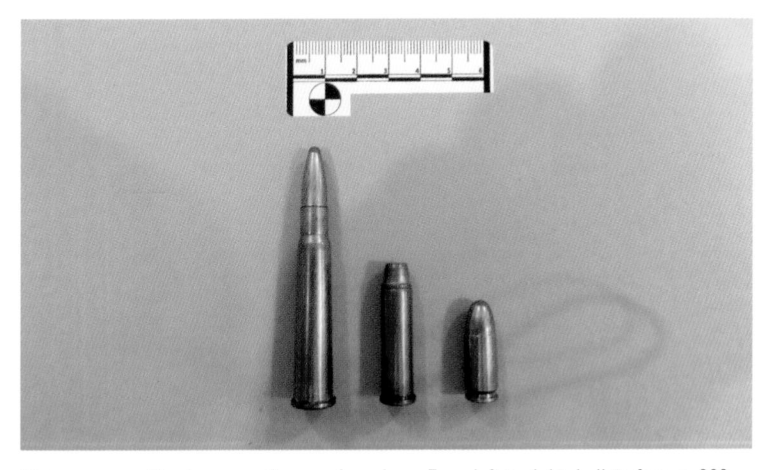

Modern ammunition incorporating powder primer. From left to right, bullets from a .303 rifle, a .38 special revolver and a 9-mm Parabellum pistol

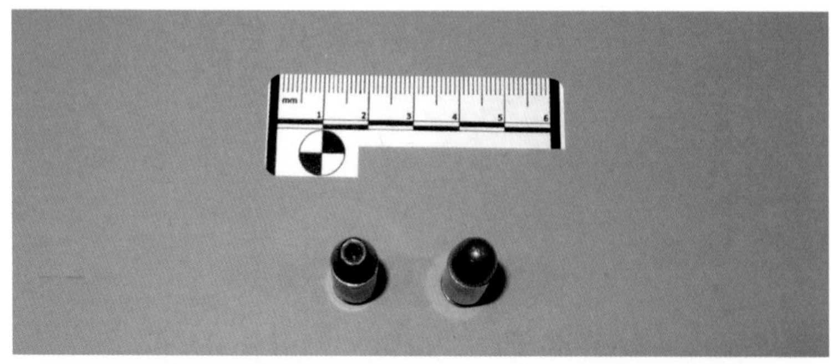

Two 9-mm Parabellum cartridges: left, with expanding bullet; right, with full metal jacket bullet

Primer in the base of the cartridge

A ballistic experiment on a pig's head ...

... and the result

copious laboratory notes concerning the validation of the new method. No notes were ever produced or mentioned.

Much later in the proceedings, it transpired that the whole process of laboratory testing had been observed by another scientist working for the prosecution. Her name was Janet Drayton, and she had opined at the time that she thought the tests only *might* have shown the presence of nitroglycerine, the major component of some explosives. In a sinister twist, she disclosed that the pages of her notebook dealing with this aspect of the test had been ripped out, not by herself.

Finally, Skuse was vague about the concentration of caustic soda, which, as mentioned, is extremely important in the final outcome of the test.

In another case, the Maguires were arrested largely because they were Irish and because of the anti-Irish paranoia that was sweeping Britain at the time. They were prosecuted after being well and truly assaulted by the police in an attempt to solicit bogus confessions. The charge against the Maguires was that they had been producing nitroglycerine-containing bombs. Again, the Crown used the work of Frank Skuse. Once more, his work was anything but accurate and definitive in obtaining the results he claimed to have obtained. Not only did Skuse give less-than-honest evidence, but the other witnesses were found by the Right Honourable Sir John May, who submitted a report on the matter to the House of Commons on 12 July 1990, to be dishonest too. The inquiry found that Crown witnesses Walter Elliot and Douglas Higgs had lied and suppressed evidence. The prosecution used thin-layer chromatography, which does not distinguish between nitroglycerine and other substances. The extraordinary response by the establishment and, in particular, Lord Denning, who was the Master of the Rolls in England, will be discussed in Chapter 17.

There is a common thread running through these cases: the propensity of Crown and state witnesses to push the available forensic science way beyond its probative envelope, always in a direction that favours the prosecution's case. In the cases of the Birmingham Six, the

Maguires and other similar cases at the time, the police used violence and threats against the detainees and their families to extract 'confessions'. Frank Skuse connived with the police and adjusted his evidence to coincide with the dates and times of the police statements. The methods used by Skuse were not validated, nor were they adequately recorded. In addition, it would appear that Skuse removed pages from his assistant's laboratory notebook. Bad science became the handmaiden to police brutality, dishonesty and judicial indifference.

Chapter 12
THE BURNING QUESTION

'We are not to simply bandage the wounds of victims beneath the wheels of injustice, we are to drive a spoke into the wheel itself.'

– Dietrich Bonhoeffer

Fires are a common occurrence in built-up areas. This is not to say that forest and open grassland fires are uncommon, but fires in urban areas account for the lion's share of loss of life and financial costs. In both cases – fires in buildings and forest or grassland fires – insurance companies have a significant interest.

In 1984, I started the first private laboratory dealing in forensic scientific matters. It developed rapidly and for a few years it was the darling of the insurance industry. Their major concern was to determine whether fires were accidental or deliberate. Huge amounts of money ride on the outcome of a fire investigation: if you insure your property, the insurance policy will not cover you if you commit arson or if you connive with someone else to burn down the property.

Fire investigation is a fascinating branch of forensic science. A fire is a chemical reaction between a fuel and oxygen in the air. It needs an ignition source as well as the fuel and oxygen. The fuel needs to be heated up until enough flammable gases are given off, and it is these flammable gases that burn, not the solid or liquid fuel itself. Once a fire starts as a result of some ignition source, the small fire must heat other flammable substances nearby to raise them to their ignition point.

The heating process can take place via three mechanisms, all of which may come into play during a fire: convection, radiation and conduction.

As the fire burns, air is drawn into the bottom of the flame. It heats up rapidly and passes through the flame – just hold your hand over a candle to feel this. The hot air passes upwards, as it is much lighter than cold air. This process, convection, is the main mechanism whereby fires spread early on in their development.

The second way in which fires heat up adjacent fuel is by radiation. This occurs when the glowing burning material gives off radiated heat, which is simply electromagnetic waves (similar to light). In fact, the heat you feel from the sun is radiated heat from all that distance away.

The third way in which heat transfer occurs is by conduction. When you place your teaspoon into a cup of hot tea, the spoon's handle will heat up in due course. (This, by the way, is a good way to establish whether your host's teaspoons are silver. Silver has a very high conductivity compared to silver plate or stainless steel, and so will heat up very quickly.)

As fires burn, they obey the rules governing these three methods of heat transfer and the progressing fire will generally move upwards and away from its origin. There are, however, a host of factors that will affect the way in which the fire burns. With experience, the observant fire investigator will be able to interpret the fire both in terms of the normal physical laws that govern fire spread and, importantly, also in terms of the modifying factors. The entire investigative process necessitates a firm foundation in practical chemistry. A working knowledge of electrical circuits and protection engineering (the study of fuses and circuit breakers) is also a useful educational background.

One of the persistent problems in the field of fire investigation is the lack of formal qualifications on the part of fire investigators. There are effectively no educational barriers to the field, which means that former firemen become fire investigators in droves. Former police operatives are another source of fire investigators; there is even one case where a police dog trainer has joined the team.

A few case examples will illustrate the problems relating to fire investigation. In 2009, a plastics recycling factory burnt down. The insurers sent two fire investigators to investigate causation. Their conclusion reads:

> Natural: There were no reports of any natural phenomena that could be considered as having contributed to the cause of the fire …
> Deliberate: This was considered as the cause of the fire as all reasonable accidental fire causes could be excluded.

Stripped of its veneer, this translates as: 'I could not find a cause and therefore I will classify this fire as deliberate.' There is an immediate presupposition that the investigators' search for the cause has been comprehensive and that the investigators were totally competent. This method of reporting is referred to as the 'negative corpus hypothesis', which is rejected by any educated fire investigator not only because it is unscientific, but because it reverses the onus of proof onto the insured.

Most modern fire investigators are familiar with the authoritative work by the National Fire Protection Association of America called *NFPA 921: Guide for Fire and Explosion Investigations*. Its authors are quite emphatic about the negative corpus hypothesis, as is John DeHaan, who writes in *Kirk's Fire Investigation*, 'It is not a great dishonor to conclude that no cause could be determined' and 'no matter how skilled an investigator may be, in the absence of adequate indicators of fire behavior or divine intervention, some fire causes simply cannot be reliably identified'.

Insurance companies do not like this because, if the cause is unknown, then the policy will have to be paid out. Considerable pressure from such companies is brought to bear on the investigator to find something untoward so that the claims manager will be able to repudiate the claim and keep the premiums.

Over the years, many 'fire investigators' have come and gone on the South African scene. I remember a venerable old fellow, Gawie, who worked for the Council for Scientific and Industrial Research (CSIR). He was the most personable and charming of men. He entered the field of insurance fire investigation some time after I established my laboratory in Johannesburg, and in many court cases our opinions were in opposition. My opponent had a number of notions about fires that were not much better than old wives' tales. He would opine that the presence of spalling on plaster – the flaking away of plaster or even surface concrete, usually caused by heat expansion – was an indication that an accelerant had been used. I was up against Gawie once in the matter of a farm fire at Otterfontein, located on the western side of Johannesburg.

In his official CSIR report on that case, Gawie started off by saying: 'As the house was situated on a farm with no provision for fire-fighting, the fire burned itself out. This complicated the investigation as the extent of destruction was such that a probable point of origin and cause of the fire could not be established.' The CSIR report then ends with the astounding conclusion that 'the fire which damaged the house at Otterfontein was caused by arson'. The reasons cited for this were the following: there was widespread spalling, there were no signs of forced entry, and an exploding gas cylinder was able to vent through the roof, which had already collapsed.

The notion that extensive spalling of plaster indicates a rapid heat build-up is just wrong. Many factors can cause spalling. In the *NFPA 921* guide, the message is very clear: 'The presence or absence of spalling at a fire scene should not in and of itself be construed as an indication of the presence or absence of liquid fuel accelerant.'

The first paragraph of Gawie's CSIR report is somewhat embarrassing if read in conjunction with the conclusion. When the report came before court in the form of an expert summary, I was astonished to see that the first paragraph had been pruned. Clearly, the purpose of the pruning was to avoid having to explain to the judge the remarkable disparity between the first paragraph and the conclusion. Apart from the melange of junk science and factually incorrect assertions in the report, the way in which it was trimmed was problematic. I was deeply uncomfortable with this and I made it my business never to act again for the attorney involved or for his advocate.

It is of more than passing interest to see what has become of the old wives' tales that made up the arson investigator's tools of trade. Spalling, as mentioned, is a useless indicator of flammable accelerants. Similarly, in the past, various investigators made use of the pattern on charred wood to indicate the speed of development of a fire. Fast-developing fires were used to infer accelerant. There is no evidence whatsoever to support the notion that charring patterns can be used to deduce speed of development or the presence of accelerant.

As far as determining the length of time that a fire has been burning by measuring the depth of char, it was always assumed to be about 2.5 centimetres per hour. Laboratory tests have shown, however, that this can actually vary from 1 centimetre to 25.4 centimetres, depending on a range of factors.

Modern furnishings are quick to combust and the temperature in a room can get to the point where the room suddenly erupts into flame. This is called 'flashover' and can occur within three minutes. Very few situations involving the use of accelerant can match the speed of normal fire development.

Crazing of glass – the appearance of fine cracks on a glass surface – used to be considered an indicator of rapid fire development. I have never found either char depth or crazing of glass to be of the slightest help in unravelling the secrets of a fire. Research has shown that crazing is caused by rapid cooling and not rapid heating.

Pool-shaped patterns on the floor have long been interpreted to mean the presence of flammable liquids on the floor. This is not so. Over the years, I have observed such patterns where there is no question of flammable liquids being present.

The age-old tradition of trying to find the origin of a fire by looking for the most badly damaged area can lead to serious errors as well. When the investigator comes onto a fire scene, he or she is presented with a final snapshot of the processes that have gone before. The development and the order of events are often not easy to discern. The area where there is greatest damage may simply reflect, among other things, greater fire load, differences in fire-fighting efforts or enhanced ventilation.

What does all this mean to the man in the street? It means that many fires have been misdiagnosed as deliberate by so-called fire investigators who are untrained, under-qualified and biased. The financial losses to the insured are incalculable.

In March 1995, a fire destroyed a Spar supermarket in Orkney, a small town on the far west of Johannesburg. A self-proclaimed fire-investigation wannabe diagnosed flammable liquids as the origin of the fire, which happened to be near the front of the shop. The fire pattern told a different tale, indicating a fire much deeper in the shop. The clincher came when I interviewed the fire-fighters, who informed me that they had stood at the very point indicated by the investigator as one of the points of origin and had fought the fire from there. So much for that being the origin.

This same intrepid fire investigator boosted his qualifications by falsely claiming to have a diploma in chemistry from a Pretoria technikon, a claim that was not reflected in the records of that institution. In fact, significant portions of his CV appeared to be false. I am aware of a number of cases in which the bogus testimony of this man wrought immense damage on businesses and lives. In the case of Piet Sandburg, for example, three investigators diagnosed the fire in question as having been caused by lightning (the case is discussed in full in *Steeped in Blood*). This man opined differently, to the evident joy of the insurer. Despite his false qualifications and dubious testimony, however, the insurance companies support and use him. The reason is not hard to fathom: he is a compliant fellow.

Recently this 'investigator acting for ABSA insurance and Mutual & Federal insurance' examined a fire in the Strand near Cape Town. He came to the conclusion that the fire was started simultaneously in four or five separate places. Normally this is a sign of arson. In this case, when I examined the scene, it was clear that the fire had started in one place on a mezzanine floor where a faulty light fitting had ignited some paper. Fortunately the insurers followed my suggestion and got a second opinion, which tallied exactly with my assessment. The claim is now in the process of being paid. But for this intervention, our intrepid 'fire investigator' would have devastated the livelihood of the owner of the business. Maybe the insurers will learn a lesson from this case.

Sometimes there is much more than money at stake. In 1987, in Texas, Ernest Ray Willis was convicted of murder by deliberately setting alight a dwelling. Two people were killed. He spent seventeen years on death row. Fortunately for him, the wheels of justice turned so slowly that he was granted a retrial before his execution took place. The fire that sent him to death row was in all likelihood accidental. The factors contributing to his original conviction were 'false or misleading forensic evidence, official misconduct and inadequate legal defence'.

In the case of Cameron Todd Willingham, the outcome was not as happy. Willingham was executed by lethal injection for allegedly setting a fire that killed his three children. After his execution, another state-appointed expert described the 'expert' testimony that had convicted him as 'hardly consistent with a scientific mindset and more characteristic of mystics and psychics'.

Part of the problem with fire investigation is that many of the investigators receive their training by way of an apprentice-type system, where the views of the senior members – the 'old wives' tales' – are uncritically absorbed by the newcomers. The other players in a fire investigation can also be problematic. The insurance companies and their lap-dog attorneys are part of the issue. One well-known firm of attorneys, through a senior partner, has gone public about its high level of trust in a particular fire investigator who has falsified his credentials and who is severely under-qualified for the job. The reason that this individual finds favour is because of his willingness to act as a tool of the industry rather than as an independent. If one works as an insurance fire consultant, it is made clear early on that the insurance companies will not brook you working against them, ever. As mentioned earlier, one of the insurance giants said to me through one of their a senior executives that if I worked against them, I would never again work for them. That sort of idiotic attitude is prevalent in the industry. It is neither intelligent nor conducive to good fire science and honest independent experts.

A major American reinsurance company has produced a document titled 'Motive, Means and Opportunity'. It is full of junk psychology and, worse, junk science, with great emphasis placed on puddle-shaped patterns, inverted-cone patterns and generally the types of hands-on practical advice now shown to be quite erroneous. This document overemphasises the usefulness of canine-detection methods for accelerants and states that 'accelerant detection canine team evidence [is] credible and persuasive'. I am of the view that unless the canine doing the pointing out is supplemented with a competent chemical analysis on GCMS, the evidence is valueless. It would be dangerous to rely on an unconfirmed identification by a dog for at least three reasons: firstly, this type of identification does not constitute valid science; secondly, one is not able to determine exactly what the dog is responding to; and, thirdly, there is no way of testing the dog's evidence by cross-examination. (And of course, the dog cannot take the oath.)

Our courts, as well as the insurance companies, lawyers and fire investigators, fail in that they often do not understand the science.

This means that bogus science put forward by an ignorant prosecutor is often not adequately challenged by an equally ignorant defence attorney, and is allowed into the legal arena by a judge who has failed to grasp the scientific issues. In the American court system, the landmark case of *Daubert* v. *Merrell Dow Pharmaceuticals Inc.* provides guidance in this regard. In accordance with *Daubert*, several factors are evaluated to exercise the gate-keeping function that should be inherent in the court's behaviour:

- Does the method generate a testable hypothesis?
- Has the method been subjected to peer review?
- Does the method have a known or calculable rate of error?
- Are there standards that control the operation of the method or technique?
- Is the method generally used and accepted?
- Has the method been established as being reliable?
- What are the qualifications of the testifying expert?
- To what non-judicial uses has the method been put?

To my knowledge, *Daubert* is fairly unique. There is no South African equivalent; it would seem that evidence is admitted on a more or less ad hoc basis in this country.

Astoundingly, when old-school-type fire investigation was challenged in another American case (*Michigan Millers Mutual Insurance Corporation* v. *Benfield*), the International Association of Arson Investigators (IAAI) acted as a friend of the court and contended that fire investigators should not be held to the same strict standards of proof expected of other science because fire investigation was 'less specific' than the kinds of scientific testing discussed in *Daubert*. That the IAAI officials could plead this kind of rubbish is difficult to believe, but is in keeping with the old tradition of fire investigation.

The need to investigate a fire scene fully must always take precedence over a swift photographic recording of the scene, with the interpretation left until the investigator returns to the office. So much is missed that way. I investigated a case where my acquaintance Gawie offered photographic evidence that a truck could not have been driven as was alleged by the owner, as he believed that the truck did not have a driveshaft fitted. Gawie's photograph showed a driveshaft lying next to the truck, but it did not show the underside of the vehicle because it was in shadow. When the photographic negative was reprinted to lighten the shadows, there in all its glory was the 'missing' driveshaft, in its proper place. The other shaft, the spare, was lying alongside the vehicle. The key to fire investigation is to photograph the points of interest that turn up in a thorough investigation of the actual scene.

When it comes down to brass tacks, the old wives' tales mean precious little. Much of what was thought of as 'fire-scene investigation' has been shown to be nothing more than mythology and junk science dressed up in overconfidence and technical terms. In a groundbreaking article in the *Richmond Journal of Law and Technology*, Thomas May deals with much of what has passed in the courts as science but which, when properly examined, has turned out to be real junk science or *ipse dixit* (Latin for 'I say so'). May points out that many fire-pattern analyses 'have long been plagued by "ill-defined and explicitly subjective criteria" based on unproven assumptions, exaggerated claims and deficient research, testing and measuring techniques'.

Until fire investigators recognise that the phenomenon we call fire is an 'uncontrolled combustion involving chemistry, thermodynamics, fluid dynamics and heat transfer', we are going to have problems. Until we have a culture of research-based investigations with adequately trained investigators, the subject will be plagued by uncertainty and poor testimony, and this will inevitably result in miscarriages of justice.

Chapter 13

ONE FOR THE ROAD

'In some circumstances, the refusal to be defeated is a refusal to be educated.'

– Margaret Halsey

There is little doubt that more than 50 per cent of all road fatalities may be ascribed to the effects of alcohol. Alcoholic beverages have been with mankind since early man discovered the relaxing effects of drinking some juice that had been left to go off in the back of the cave. Noah discovered the effects of excess in the most shameful way, and his children's observance of the old man in his cups had (if we are to believe the scriptures) most unfortunate consequences for his son, Ham, who got himself into serious trouble.

Anyway, alcohol has been around since the beginnings of recorded history and before. The correct chemical name for alcohol is ethyl alcohol. It is a short-chained alcohol with only two carbons in the chain. Its structure is:

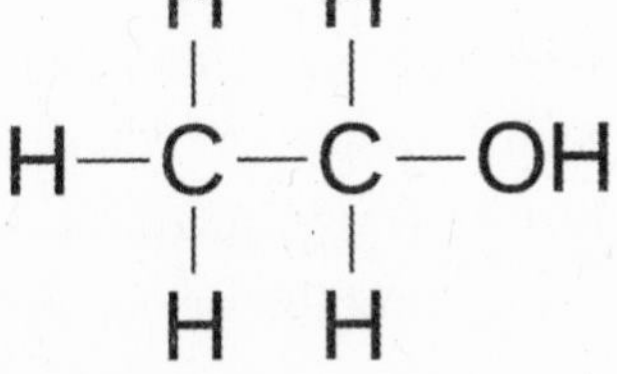

```
    H   H
    |   |
H — C — C — OH
    |   |
    H   H
```

Most, if not all, commercially available alcohol is produced now as in times of yore, by allowing a micro-organism such as yeast to convert the naturally occurring sugar in fruit (grapes, apples, juniper berries), grain (wheat, barley), potatoes and a host of other substances into alcohol. Yeast uses sugar as an energy source. Through a simple pathway called glycolysis (Greek for 'splitting of sugar') or, eponymously, the Embden–Meyerhof pathway, which encompasses a series of eight chemical reactions, each catalysed by a specific enzyme, the glucose is converted to much shorter chain compounds, one of which can be rapidly converted to ethyl alcohol. Other short-chain compounds that can be formed in the same process include lactic acid, which is painfully well known to all those who engage in strenuous physical exercise, and vinegar (acetic acid). Humans do not possess the required enzymes to convert the used products of glycolysis into alcohol, but a wide range of micro-organisms, including yeasts and other common microbes, do have the necessary enzymes.

Generally, most alcoholic beverages are produced by allowing yeasts to convert natural sugars into ethyl alcohol. This simple process does not produce a very concentrated solution of ethyl alcohol because alcohol at higher concentrations starts to inhibit the fermentation process. So mankind, in its inexorable forward march, started to take the weaker alcoholic solutions and to distil them so that the flavourants and the more volatile ethanol would produce a much more concentrated alcoholic solution in the distillate. In this way, liquids like wine and beer-type drinks were converted into brandy and whisky and the numerous other delectable potions found at your local liquor outlet.

When an individual drinks a solution containing alcohol, it passes first into the stomach, where very little of the alcohol is absorbed. Thereafter, it moves into the first part of the small intestine, the duodenum, where most of the absorption takes place and the ethanol moves into the bloodstream. From there it is conveyed to all parts of the body and, most importantly for our purposes, to the brain.

Because of the small size of the ethanol molecule and its particular properties, it has no difficulty in passing the blood-brain barrier, and this is where the interesting and sometimes unfortunate effects of alcohol take place.

Alcohol is a fairly mild central nervous system depressant at low concentrations, but at high concentrations in the bloodstream it will produce coma and death. At lower levels, it produces some gentle physiological effects, such as a flushed face. Its depressant effect on the central nervous system produces a level of euphoria and some loss of restraint – in other words, often the person we see at a party as a jolly good fellow, the life and soul of the party. The effects of ethanol go much further than dampening your inhibitions, however. One of the first things to suffer impairment is fine motor control. The ability to perform and coordinate multiple tasks at the same time is especially negatively affected. A person's reaction time is also diminished, although it must be said that this occurs only at higher levels of blood-alcohol concentration. At levels of 0.1 grams (percentage of alcohol), most people exhibit a marked incoordination of skilled movements and difficulties in maintaining balance. Even at these low levels of blood alcohol, the individual's ability to drive a car safely may – and probably is – significantly impaired. The most important area is in dual-task performance and divided-attention performances. Tracking performance – the ability to follow moving objects, such as the car in front of you, and to deal with the road moving in relation to the car – is significantly impaired, as it requires rapid small adjustments in the muscles of the hands and eyes, as well as a high level of hand-eye coordination.

The time taken between drinking and absorption means that there is a slowly rising blood-alcohol concentration after drinking – the so-called absorption phase. Thereafter, there is a plateau and the blood-alcohol concentration slowly drops. The complete blood-alcohol curve looks like this:

Issues in Alcohol Testing

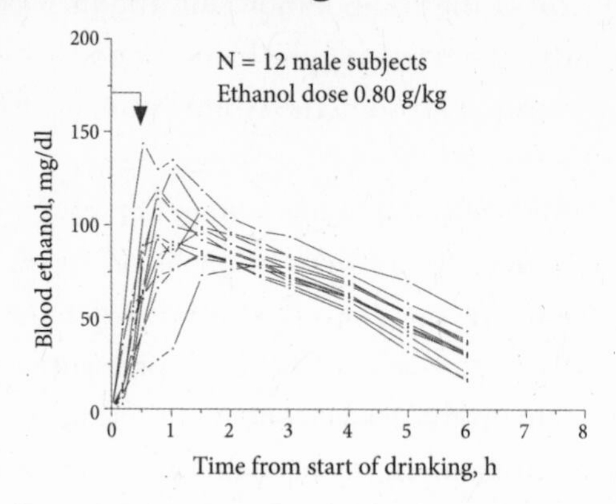

Individual concentration-time profile of ethanol in venous blood for 12 healthy men who drank 0.80 g ethanol per kg body weight in 30 min after and overnight fast

Source: Alan Wayne Jones and Derrick J. Pounder, *Forensic Issues in Alcohol Testing*

From this graph it can clearly be seen that the peak alcohol concentration occurs after approximately forty minutes. It is also quite obvious that there will be occasions where the individual's blood-alcohol levels will be exactly the same for a short moment in time; one on the upward side (absorption phase) of the curve and another on the downward side (excretion phase) of the curve. It is also well known that, for a given concentration of alcohol in the blood, one is more inebriated on the upward side than the downward side. This is known as the 'Mellanby effect'.

Part of the massive death toll on the roads may significantly be placed at the door of alcohol. It is thus an important part of any road-safety campaign to investigate and prosecute offenders who are driving while under the influence of alcohol or drugs.

The gold standard test for measuring blood-alcohol concentration has for many years been an instrument called the gas chromatograph, described in previous chapters. This method is capable of being automated and it replaced earlier, tedious chemical methods. The standard practice in state laboratories to measure blood alcohol is to split the sample and to run it on two separate instruments so that the results can be checked against each other. The system requires that the two results are not to differ by more than 5 per cent, and the lower level is always used.

The quality of staff in state laboratories has unfortunately plummeted. The difference between the two analyses is now set at 10 per cent. Analysts who were expected to complete between twelve and fifteen toxicology analyses per month have dropped the output to about three, and even these are badly done. It now takes approximately five to eight years to get a toxicology result back from these laboratories. This means that any post-mortem requiring an answer on possible poisoning cannot be finalised within that time. The effect on insurance claims and on winding up estates is incalculable.

Blood-alcohol results that should be available at the first remand (about three months) are delayed for about two years. When the results finally become available, they are so bad that they will not often pass legal muster. The duplicates, as mentioned, should not differ by more than 5 per cent; I regularly see results that differ by about 15 to 18 per cent and, in one instance, a sample was analysed four times with a spread of results differing by 100 per cent.

The reason for this is not hard to establish. The staffing policy has changed to allow only the appointment of previously disadvantaged individuals. The educational level of the staff is hovering around the diploma level and, when asked to do an elementary test on basic chemistry, the incoming students fail hopelessly.

The head of the blood-alcohol section of the Cape Town Health Chemical Laboratory was taken apart in a court case (*Larg Heinke* v. *Andrew Hubbard*). The state analyst in this matter was Mr Bongile Lengisi, who was at the time the chief forensic analyst. His evidence was so poor, his knowledge of the subject so deficient, that in the face of a defence put up by Dr Neels Viljoen, Lengisi took refuge in denial and simply refused to accept the obvious. For instance, he would not accept that an incorrect calibration 'will or may affect the values calculated as part of the analytical process'. This is so basic that Lengisi's evidence was roundly rejected in this matter. It seems that the punishment for such incompetence is promotion. Lengisi is now the head of the national Department of Health's Forensic Chemistry Laboratories in Johannesburg.

This abysmal situation pertains across the board in the chemical laboratories, which are also expected to be the watchdogs of imported chemicals. Some years ago, a vast consignment of Chinese fertiliser was procured and used in northern KwaZulu-Natal. The laboratories' failure to analyse this adequately, or at all, resulted in the pineapple-growing area being poisoned with high levels of cadmium. This poisonous import not only flattened the premier pineapple-growing industry, but it also went into pig feed and chicken feed, with disastrous consequences. Another example of the failure to analyse imports (again from China) resulted in melamine ending up in milk imports. Melamine is added by unscrupulous Chinese businessmen to artificially boost the protein levels of the product. It is basically a waste material, but in chemical tests it responds in the same way as a protein does. Unfortunately, it is highly toxic to the kidneys. The net result of this was that many dogs who consumed the feed made from this toxic brew had to be put down.

The situation in the chemical laboratories has deteriorated to the point that they are essentially dysfunctional. In Cape Town, the ANC government was ousted by the Democratic Alliance in 2009. The new provincial Minister of Transport, Robin Carlisle, set himself a target of halving the road deaths during his tenure. One evening during a talk that I attended, Carlisle outlined his path forward. He noted that the laboratories were dysfunctional and indicated that they would therefore bypass the laboratories by making use of an instrument called the Dräger breathalyser. The actual instrument to be used was the Alcotest 7110 Mark III. Curiously, various traffic departments throughout the country had been given this equipment by none other than South African Breweries (SAB). I am not sure what deranged logic prompted this donation, as SAB spends millions on advertising designed, presumably, to get people to drink more. For them to provide the means to have their clients locked up seems to me to be running with the hares and hunting with the hounds at best, and hypocritical at worst. In any event, Carlisle intended to rid the roads of drunken drivers by using this machine as the primary – and only – evidence in the road campaign.

At the end of the talk, I approached Carlisle and suggested that we discuss the matter before the campaign was launched. This duly happened at my offices in early 2010. Advocate Paul Hoffman and I met with Carlisle and pointed out the well-known and legally dangerous difficulties with the instrument.

The consequences of a criminal conviction are serious and can have life-changing effects on the person convicted. Some convictions may result in a custodial sentence and, in South African prisons, with the high prevalence of HIV and AIDS, this is the equivalent of a death sentence. So, in the light of all this, the measurement of alcohol levels in the suspect must be as good as, if not more accurate than, the gold standard of using blood and gas chromatography.

Yet the Dräger breathalyser is beset with problems. Not only is the instrument not specific for ethanol, but the results are extremely variable and are dependent on lung capacity and the breathing pattern of the person taking the test, among other things. The whole relationship between blood-alcohol level and breath alcohol is very variable.

When Hoffman and I pointed out all of this to Carlisle, he agreed to a further meeting at his offices in the Cape Town city centre. Present at this meeting were Advocate Christinus van der Vijver from the National Prosecuting Authority and two lackeys from SAB. The meeting ended with the agreement that a Dräger representative be present at a follow-up meeting. Rob Brown, the local Dräger representative, had been present at the first meeting, but he, unfortunately, did not know how the instrument worked.

I duly presented myself at the next meeting, to be given the information that the Director of Public Prosecutions, Advocate R.J. de Kock, had sent Carlisle a letter that is almost majestic in its failure to comprehend the situation. I have included a copy of this letter as an appendix for readers' amusement, together with my reply (see Appendix C). The net effect of De Kock's letter resulted in my being unceremoniously ejected from the meeting.

There the matter would have rested but for my eye alighting on an article in the Cape Argus informing me that the state intended to prosecute a young man for drunken driving in the High Court using the Dräger instrument only. Legal Aid was to take on his case, and on the state side were Van der Vijver and Billy Downer, both very experienced senior counsel. Here, indeed, was a horse race with two thoroughbreds pitted against cart horses. The whole case, State v. Hendricks, was sneaky in that Legal Aid had no expert. Indeed, as it transpired in court, its representatives were less than adequate in legal ability. Were the state to prevail in this case, the judgment would be binding on all lower courts, which is where the vast majority of drunken-driving matters are heard. In other words, this was a barefaced effort by the state to force the Dräger down the throats of the public, knowing full well that the instrument was not up to the legal requirements expected of it.

I felt that justice was not being served by downgrading the standards by which alcohol was measured. Much better to fix the problem at the laboratory and return to the gold standard of blood testing. I started by utilising the help of William King, one of the most experienced criminal advocates at the Cape bar. Next in line was Paul Hoffman, who, as director of the Institute for Accountability in South Africa, had more than a passing interest. A short while later, I encountered Advocate Derek Mitchell. On hearing of our quest, he volunteered to become our leader. In my search for scientists to help me with some of the technical issues, I was granted the great good fortune of enlisting the help of Professor Andrew Wilkinson, an associate professor of electrical engineering at UCT. He came with just the right sort of knowledge that we needed to see into the dark little soul of the Dräger black box.

The first of the dirty tricks attempted by the prosecution was to deny the defence team access to the source codes of the instrument.

These codes are basically the internal operating instructions programmed into the machine. They reveal the thinking of the designer as well as the internal algorithms, or instructions that operate the machine, and form the very heart of the instrument. Without this information, we were faced with a box that simply spat out an answer, and there was no way of checking this answer. Downer and Van der Vijver kicked off by claiming that the information we sought was patented and privileged. That was just so much nonsense. Both prosecutors knew – or should have known – that the accused is entitled in our law to know the basis of the prosecution. Without that we are in contravention of the common law as well as the Constitution. When I asked our leader to inform the prosecutors that we would launch an immediate application to compel this disclosure, they backed down and released it – by allowing Andrew Wilkinson to go to their offices and to read it there; in other words, to make it as inconvenient for the defence as possible. The next prosecutorial trick was to make the source codes available only in German. Unfortunately for them, Wilkinson is fluent in German, having studied in Germany at a postgraduate level. With access to the source codes, a whole new vista opened up for the defence team. We discovered a number of fundamental problems with the instrument.

As mentioned previously, the blood-alcohol test is conducted in duplicate. Dräger boasts that its instrument can do the same. There are two different types of sensors in the instrument. One depends on the fact that, if a beam of infrared light is passed through a vapour containing ethanol, its intensity will be reduced. A second device within the instrument consists of an electrochemical cell, which generates an electric current in the presence of ethanol; the more ethanol, the more current. The two devices are coupled in such a way that if the reading by the one differs from the reading by the other, the analysis is aborted. However, when we examined the source code, it was apparent that the difference between the two devices could be as large as 40 per cent without the instrument noticing it –

a far cry from the 5 per cent difference in duplicates required in a blood test.

The second feature of importance was that only one of the detectors in the instrument was calibrated. Technical problems with calibration of the fuel-cell component meant that the instrument was functioning on only one detector that could be relied on.

Prior to *State* v. *Hendricks* I had arranged a test. About thirty-six people from my circle of friends were invited to a party at my house. They were all given normal party fare, with wine and beer. Each person had agreed to give a blood sample and to be simultaneously tested for breath alcohol. The relationship between the blood level and the breath level was then expressed as a ratio. One might expect that there would be some kind of predictable constant ratio. This was, however, not so. The ratio was highly idiosyncratic, and even in the same individual it varied depending on whether the individual was in the absorption phase or the excretory phase. All of this information was passed on to Carlisle, to no avail. The findings of our test were in accordance with the internationally published results that were researched in preparation for the trial.

To state that the trial ended in disaster for the prosecution is putting it mildly. During the trial we were able to show that the instrument provided by SAB was the cheapest model. It took no cognisance of the temperature of the breath (vital for the accurate operation of the instrument). In addition, it is clear from elementary pulmonary physiology that Henry's law, the chemical law underpinning this instrument, is not applicable to expired air. Henry's law states that if one has a solution of a volatile compound in a closed container and one allows the solution to reach equilibrium, the concentration of the volatile substance in the air above the liquid in the container will be proportional to the concentration of the volatile substance dissolved in the liquid. In this context, it is being applied on the basis that the concentration of alcohol in the blood will allow a certain amount to escape into the air passing into the lungs. That air, according to the theory, will be at equilibrium with the blood, and the concentration of alcohol in the air will be a measure of the concentration of alcohol in the blood of the individual. Unfortunately, the air in the lungs is never at equilibrium. It is not a closed system and it is not at a constant temperature. Therefore the provisions of Henry's law are not fulfilled and it cannot be applied to this particular situation.

In further research done pursuant to this case, the technical legal team for the defence found that the Dräger breathalyser did not even comply with the South African Bureau of Standards' specification for such equipment. One might have expected that two prosecutors of the age and experience of Downer and Van der Vijver would have ascertained that elementary legal point before their hasty and devious attempt to push through this unsuitable instrument.

Not only was the instrument legally unfit for service and physiologically inaccurate, but several of its much-vaunted claims were shown to be false in evidence given by state witnesses. Peter Berman, who was called by the state, proceeded to shoot them in the foot by proving that mouth alcohol would give a false reading, in contrast to the claims made by Dräger itself. A thorough consultation with the state witnesses might have alerted Downer and Van der Vijver to this.

The evidence of Dr Jürgen Sohege, the Dräger expert who was called for the state all the way from Germany, indicated that Dräger itself suggested that certain additional safeguards be incorporated into the test, namely that at least two consecutive tests be performed with a suitable separation time in between. The traffic official interviewed by Wilkinson and me before the trial was quite oblivious to this requirement. When asked if she did two measurements, she looked at us in wild surmise and said, 'Why, what for?'

During our research into the ways of breath-alcohol testing, we also found that the results were posture-dependent as well as reliant on the breathing pattern of the person taking the test.

It is of some interest that the instrument bought by SAB for use in South Africa is not acceptable for use in its home country. In Germany a much more sophisticated instrument it used. It incorporates some of the safeguards that we identified as being necessary for use in a local prosecution. It would seem that SAB were trying to look good on the cheap.

Interestingly, the Dräger has had mixed receptions in different countries. In Germany, the upgraded model of the instrument is in use as a standalone for so-called administrative offences, which carry a penalty of a fine only. For more serious offences, the German authorities require a further blood test and do not accept the Dräger as a standalone piece of evidence.

In Australia, the road traffic offence is essentially failing the breathalyser test. In other words, if the Dräger result places you above the legal limit, that is enough; it does not mean anything to the authorities that your blood alcohol (which is what really matters) is below the legal limit. Given the failings of Dräger, this is not good law – the concept is not tethered in reality. The profound implications of this sort of law are discussed in an *Australian Journal of Forensic Sciences* article by Ian R. Coyle, David Field and Graham A. Starmer, titled 'An Inconvenient Truth: Legal Implications of Errors in Breath Alcohol Analysis Arising from Statistical Uncertainty'. The article argues that the accuracy of the breathalyser overestimates the blood-alcohol levels and that the assumptions on which the functioning of the instrument are based are fundamentally unsound. For the Australian legal system to trump a scientific truth with bureaucratic fiat is as stupid as one of the American states trying to legislate pi (π) as a rational number, which it is not.

In the US, each state differs in its acceptance of the instrument. Some allow it as standalone evidence and some require an additional blood test. In 'Of Black Boxes, Instruments and Experts: Testing the Validity of Forensic Science', Jennifer Mnookin discusses the Dräger 7110 with regard to the leading case *State* v. *Chun*. While I agree to a large extent with the very perceptive writings of Mnookin, I must take issue with her stance on the breathalyser. She is correct in that, to a great extent, the breathalyser result is often wholly dispositive of the matter in question. Thus, we have the unsatisfactory situation where the ultimate issue is decided by a computer programmer who is not available to be cross-examined. With great deference to Mnookin, input/output testing will not ever be a sufficient safeguard. In other words, the functioning of the instrument is validated by putting samples of known concentration in one end and checking the results at the other end. If they correspond, then the instrument passes muster. This approach will not detect all the other areas of deficiency in the functioning of the instrument, such as the discrepancy between the detectors.

In *State* v. *Hendricks* it became clear that the safeguards alleged by Dräger simply were not operational in the equipment under discussion. This revelation would never have seen the light of day had we been denied access to the source codes. The disabling of the temperature-detection function, too, became visible only when we examined the source codes. Various operator influences on the outcome require that the court should have a deeper insight into the function of the instrument than simply a blind faith in the input/output sufficiency. Such an instrumentalist approach for the breathalyser is bad, but if this precedent were to be set in other areas of forensic science, the outcome would be catastrophic.

Interestingly, the position taken by Dräger in *State* v. *Chun* was the exact position taken by the company in *State* v. *Hendricks*, namely to try to avoid disclosing the source codes of the instrument. The New Jersey court in *State* v. *Chun* ordered that the source codes be disclosed. It is my considered opinion that the American experts who examined these did not do so in sufficient detail. They failed to uncover the inconvenient gaps in the software design that Wilkinson uncovered in *State* v. *Hendricks*. It was of some significance that Dr Sohege conceded every point that Wilkinson had exposed during his lengthy cross-examination by Derek Mitchell.

The result of this court case has been disastrous for the state. Had they used the breathalyser in the lower courts, in the normal way, some would have succeeded and some would have been challenged, but nothing would have been binding. It would have been business as usual. By forcing a High Court decision that went against the prosecution, the state is prevented from using this equipment until it has been modified and improved, and until it can pass the next High Court challenge. The immense stupidity of embarking on this course cannot be overstated. The state and the local government were warned and informed of the problems. Carlisle thought that his particular brand of political bluster would trump science. It did not.

The real tragedy of this whole circus is that, had the state players taken advice from people who were not trying to sell them equipment, they would have found themselves in a different place. The laboratory measurement of blood alcohol can be automated and done with a turnaround time of less than forty-eight hours. The available technology allows for a mobile analytical facility that could be parked at hotspots, such as outside clubs or sports stadiums, to check alcohol levels. Drug levels can also be measured and identified. But this has not happened. The extent of the final disaster caused by our government agents' actions is hard to overestimate.

Chapter 14
THE WAR ON DRUGS

'To say that the first casualty of war is truth is to miss the rather more important point that a principal weapon of war is lies.'

– Harry M. Collins

Drugs with a narcotic effect are not a new phenomenon. Morphine takes its name from the Greek mythological god of dreams, Morpheus. It is one of the active components in the juice of opium poppy (*Papaver somniferum*) extract. The purified component was isolated from opium in 1804 by a German chemist, Friedrich Sertürner. It became commercially available in about 1817 and Merck, who knew a good thing when they saw it, obtained the rights to sell it in 1827.

Morphine was, and still is, the most remarkable drug for extreme pain relief. It was for many years virtually the only effective drug available for pain relief. Unfortunately, morphine has, together with many of its close relatives, a marked tendency to produce addiction as well as to become less effective at a given dose, requiring increased doses for a continued pain-stilling effect. Addicts rapidly become tolerant to the drug and eventually the doses required become toxic.

Opium, the raw material from which morphine is extracted, had been used in China since at least the seventh century. In the seventeenth and eighteenth centuries, British trade with China was producing a serious deficit in the British coffers. The Chinese would accept payment in silver only and the British had to purchase silver, as they had moved on to the gold standard. Not surprisingly, as the British demand for China's tea increased, the British, who had been importing opium from India for a while, started to trade with the Chinese in opium. At the time, the use of opium in China for pain control was limited, and the substance was not used recreationally. Because of its addictive nature, it became an instant solution to the silver trade deficit for the British. Ultimately, attempts by the Chinese authorities to stop this harmful practice resulted in hostilities breaking out between Britain and China. Generally, however, the import of opium from the British was unstoppable.

Morphine is just one of many drugs with a narcotic effect. There are well over 100 substances that fall under the title narcotic. One of the most addictive is a drug called Mandrax (active ingredient methaqualone), which is one of the most successful sleeping pills ever produced. Unfortunately, it too was discovered to be hopelessly addictive. Our local South African addicts have made a significant contribution to international pharmacology research by mixing methaqualone with marijuana and smoking the concoction. This is known as a 'white pipe' and is extraordinarily addictive.

Slowness to respond to the knowledge of various drugs' addictiveness has been problematic. Morphine addiction became a major problem after the American Civil War, with many Northern and Confederate soldiers becoming hooked on it in their efforts to control pain resulting from injuries. In the early 1870s, a new 'wonder drug' was synthetically prepared, trademarked and patented. The drug was originally synthesised by C.R. Alder Wright in England at St Mary's Hospital. The giant German drug company Bayer marketed the drug under the trade name 'Heroin', from the Greek *Heros*. They thought that the drug would have a heroic effect by rescuing morphine addicts.

Unfortunately, heroin is just as addictive as morphine, if not more so, and this was not Bayer's finest hour. Below is an early advertisement indicating how this harmful drug was used for trivial conditions before it was adequately tested.

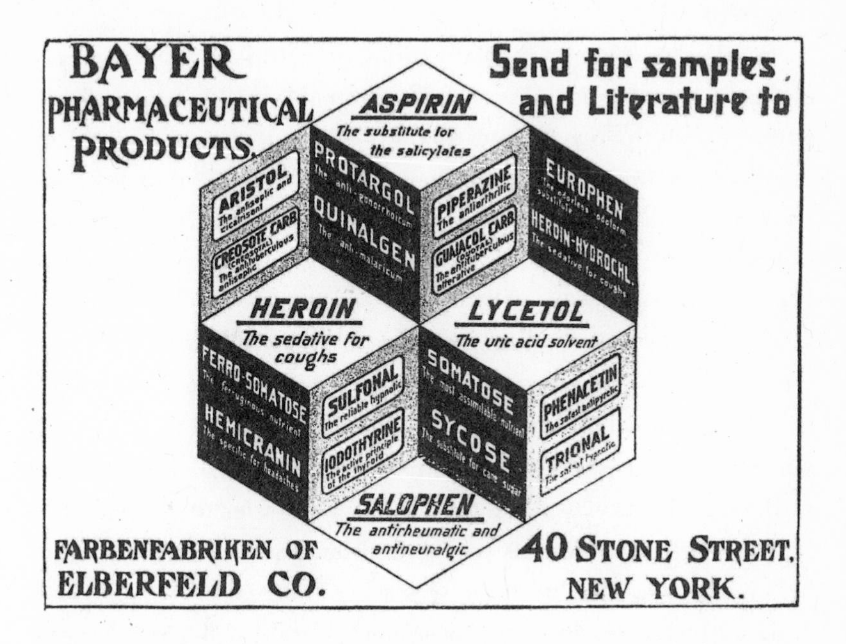

Cocaine is another exceptionally addictive drug. Obtained from the leaves of the coca plant, it has the clinical effect of making you feel alert and on top of the world – nothing is beyond your ability. It also engenders enhanced sexuality. Sigmund Freud, better known for his theories on psychoanalysis and whose lurid theories on sexuality are still discussed in some circles, came perilously close to becoming a cocaine addict. He had come across a paper by a German army surgeon on the use of cocaine for treating exhaustion in soldiers. Freud wrote what he himself described as a 'song of praise' to this magical substance and hotly denied that it was in any way addictive.

'Cocaine will make the coward brave, the silent eloquent and free victims of alcohol and opium from their bondage.' So read an advertisement from the pharmaceutical giant Parke-Davis in the early part of the twentieth century. Cocaine's origins can still be seen in the name of Coca-Cola, which was invented as a temperance drink by pharmacist John S. Pemberton. The original recipe contained extract from the coca plant, and it was only in 1929 that the product was reformulated.

Laudanum was an alcoholic solution of opium containing several of the opium alkaloids. It was easily obtainable and widely available without prescription. In fact, the famous Ed Ricketts (depicted in Steinbeck's *Cannery Row*, mentioned earlier), who had on his own admission no licence to practise medicine, was dishing out laudanum to Lee Chong's wife. (Lee Chong owned the general dealer store in Cannery Row in Monterey.)

Drugs have therefore been around for an awfully long time and, with the belated awareness of their negative side effects, they have become ingrained in much of our culture. Even drugs such as LSD, which was synthesised by Albert Hofmann at Sandoz Laboratories in Basel in 1938, was marketed for a while under the trade name Delysid, before its serious and sometimes permanent side effects became known. This, however, was never a success.

In a much more sinister project, the Central Intelligence Agency (CIA) launched an operation in the 1950s called MK-Ultra, where prostitutes were used to lure businessmen to brothels, where they were secretly dosed with LSD and then placed under observance. How low can you go? If you think this is beyond the pale, a reading of Gordon Thomas's book *Secrets and Lies: A History of CIA Mind Control and Germ Warfare* will convince you otherwise.

Matters were not made any better by the widespread use of the drug Benzedrine during World War II to keep fighting forces alert and awake during periods when sleep was not advisable. This drug is highly addictive and rapidly spawned an epidemic of addiction. Addiction is such a powerful human urge that it is almost impossible to cure. For most narcotic addicts, the rehabilitation rate is in the low single figures, and the relapse rate is high. A vast amount of time and police resources are spent worldwide in combating this scourge and, in the end, there is little to show for the effort.

We can therefore take it that mankind's flirtation with dangerous drugs through medical ignorance, misprescription and government stupidity has blossomed into a consummate marriage. The authorities, while prohibiting drugs in some cases and limiting their availability in others, have freely made use of various drugs for their own purposes. The CIA has used a variety of drugs in the pursuit of mind control, and our own Lothar Neethling experimented with Mandrax to achieve crowd control in the late 1980s.

Various authorities have conceded to me that, in practical terms, the war on drugs in South Africa is being lost. Rehabilitation clinics are full and waiting lists are long. The level of police corruption in the supply chain is significant. In Durban, the head of the local drug-enforcement arm of the police was himself involved in selling drugs to the underworld. Dr Liza Grobler, one of the leading criminologists in South Africa, has recently written a book called *Crossing the Line*, a depressing catalogue of police involvement in serious crime. She lists how police officers do route clearance for gangsters and how they provide escorts for shipments of illegal goods. A gang boss surmised that 30 out of every 100 police officers are corrupt.

Let us examine the typical prosecution for possession of, or dealing in, drugs. Intelligence must be obtained as to the whereabouts of a cache of drugs, the dealing of drugs or the arrival of a shipment of drugs. A raid or an interruption must be planned. This leads to significant problems because corrupt cops often tip off the criminals. It should not be forgotten that Jackie Selebi, who was at one time our Commissioner of Police and the head of Interpol, was a close associate of a convicted drug dealer who did Selebi many financial favours.

If the drug raid or interruption is successful, the consignment must be seized and documented. In some instances, the actual weight of drugs seized can be up to 1 000 kilograms or more. Clearly it is neither practical nor feasible to test the whole consignment. Samples must be selected and they must be representative. They then need to be analysed. In previous times, analysts would make use of thin-layer chromatography, one of the first 'modern' methods of separation in analytical chemistry. In its most basic form, a drop of the liquid or solution to be analysed is placed onto a silica plate. The plate consists of a glass backing, onto which a very thin layer of silica powder has been coated. The plate with the material to be analysed is placed in a tank containing a suitable solvent at the bottom. The solvent moves up the plate by capillary action, rather like a candle or lamp wick. As the solvent flows over the spots on the plate, the moving solvent carries the components in the mixture along with it. Different substances have different mobilities in this system and so the compounds will separate on the plate by reaching different levels with respect to the solvent front. The distance migrated may be measured, and is a crude method for identifying the substance; different compounds migrate differently on the plate.

The next step in the identification process is to make use of gas chromatography, which was discussed briefly earlier. This is a much more sophisticated variation of thin-layer chromatography. It differs from thin-layer chromatography in that the mixture to be analysed is injected into a thin tube and the flow of an inert gas carries the mixture through the tube, the different compounds of the mixture emerging separately at the other end. Various detectors can be attached to the gas chromatograph, but the most useful is a mass spectrometer, which takes the sample as it emerges from the gas chromatograph and renders it into a pattern that can be used to identify the material with a high level of confidence. Thus, in today's world, gas chromatography-mass spectrometry (GC-MS) is an entry-level method for drug and other forensic analysis.

Whatever transpires at the laboratory level is entirely dependent on the groundwork done at the scene of the crime. The crime-scene officers that I have come across are less than competent. In the case of the Lotz murder in Stellenbosch, as many as eight to ten unauthorised policemen were inside the deceased's flat without protective gear and without any reason to be there other than prurient curiosity. This level of incompetence is not limited to a murder scene.

As far as drugs, drug supply and the attempts to curtail the use of drugs go, there are several factors that militate against success, the first, and most important in my opinion, being the immense addictive potential. The second factor is the vast amount of money that can be made from trafficking drugs. Some years ago, I was involved in a case where the accused had manufactured methaqualone worth R186 million. By today's values, that amount translates into the billions. The third problem stems from the first and second: our authorities are easily corruptible. This is not surprising, given the vast amounts of money involved and the poor remuneration of the police. I have been in personal contact with the South African Medical Research Council (MRC) and the message is the same: they do not see the war on drugs being won. In fact, there is some ambivalence about what needs to be done.

Professor Charles Parry, who is the director of the Alcohol and Drug Abuse Research Unit of the MRC, has written extensively about the problem. In a recent publication, *Harm Reduction in Substance Use and High-Risk Behaviour: International Policy and Practice*, he notes that 'the emerging pattern in recent years appears to be one of increasing heroin, cocaine and amphetamine-type stimulants ... Treatment demand for heroin misuse in South Africa increased sevenfold compared to other substances over an eight-year period from 1998.' To give some indication of the scale, the value of the drugs in trade in Guinea-Bissau, consisting mainly of South American cocaine moving to Europe and the US, was estimated as the equivalent of the country's entire national income.

The HIV/AIDS pandemic has also been seen to be exacerbated by drug use, in particular the increase in injected drug abuse. Needle-sharing is rife, with 86 per cent of users having shared a needle at least once. It is also likely that drug abuse leads to risky sexual behaviour, with all the attendant consequences.

Despite increased resources and a harsher approach, the attempts to curtail drug use are to little or no avail. It is short-sighted to think that a prison sentence will have any effect. If the offender is not an addict when he goes into prison, he almost certainly will be when he comes out: drugs are freely available in our prisons – yet another opportunity for corrupt officials. Furthermore, the prisons are so destructive of the entire fabric of an individual's self-esteem that drug addiction is unlikely to cease. To continue spending vast sums of money on the same old failed behaviour is not intelligent; there is a need for a radical new approach.

It is my view that the driving force behind drug addiction is the profit motive. I have little doubt that if you reduce or, better still, eliminate the money to be made, the trade will largely collapse. How is this to be done? Firstly, each and every drug is very cheap to make. Morphine, for example, which is not expensive, is easily converted into heroin, and the street value of a kilogram of heroin is vast. The cost to produce it would be in the region of R15 000. If the government were to set up or commission a pharmaceutical company to produce pure drugs, the price of production would drop even further. These drugs could then be made available to addicts free of charge, together with sterile syringes and needles. Crazy, you may say, but consider this: drugs of abuse are being used in ever-increasing amounts. The efforts to stop this have proved fruitless. Part, if not most, of the problem lies in aggressive recruitment by dealers and pushers who are driven by profit. If the drugs were available for free, the motive to recruit would fall away. Since drug-pushing starts in children who are still in primary school, the number of children who become involved would dwindle rapidly, as penalties for pushing and recruitment would remain. It should also be remembered that entry into the world of drug addiction is free. Once the victim is hooked, the price tag appears.

Secondly, the adulteration of the end-user drugs, with all manner of harmful excipients – those substances used as a diluent or vehicle for a drug – would cease to be a problem. I was involved in a case where cocaine used by a young man was contaminated with strychnine. The pharmaceutical-grade drugs would be of a proper grade, not the highly impure active ingredients produced by clandestine laboratories.

The problems associated with the use of shared needles would also vanish. The financial motive for corrupting police and other authorities would cease. The recruitment of new addicts would stop. The need for petty crime to feed the addicts' need would vanish, so the crime rate would drop. The courts would be freed up to deal with more serious crime and valuable resources could be redeployed. I think that if these measures were to be implemented, we would get to grips with the drug problem. Tentative moves in this direction are faintly suggested in the work of Charles Parry and others.

Whether or not there will be an official with the courage to adopt these ideas, or measures like them, remains to be seen. In the meantime, we shoot ourselves in the foot, again and again. Expending court time and police resources to track down and prosecute people is futile: we are largely prosecuting people for an addiction, and we should regard addiction as a medical condition, not as a criminal activity. Pursuing and punishing the pushers and dealers would be a much greater contribution to justice.

Chapter 15

BITING THE BULLET

'At the heart of science is an essential balance between two seemingly contradictory attitudes – an openness to new ideas, no matter how bizarre or counterintuitive they may be, and the most ruthless sceptical scrutiny of all ideas, old and new. This is how deep truths are winnowed from deep nonsense.'

– Carl Sagan

'Til the infallibility of human judgements shall have been proved to me, I shall demand the abolition of the penalty of death.'

– Marquis de Sade

The development of firearms over the past 400 years or so has seen unrivalled amounts of human effort and ingenuity invested in it. The early weapons were muzzle-loaders, which meant that gunpowder needed to be introduced through the front end of the gun, followed by the bullet or, in the case of larger guns, the cannon ball. Interestingly, several English phrases in common use today derive from early guns: flash in the pan; lock, stock and barrel; to go off half-cocked, and so on.

The weapon was fired by igniting the main charge through a touch hole, thus igniting the gunpowder within the barrel. The early weapons were effectively hollow tubes that had a muzzle end and touch hole at the other end. They suffered from many defects and disadvantages: in wet weather the igniting fire was hard to keep alight, and having to handle the fuse was detrimental to the two-handed handling of the gun, so they were clumsy and difficult to aim accurately. The loose fit of the projectile in the barrel also produced problems, in that the hot pressurised gases from the burning propellant could leak past the bullet and would result in lower muzzle velocity and diminished effectiveness of the projectile on the target. To improve the seal and to prevent leakage, the bullet was rammed down the barrel with either paper or a cloth around it to make the whole lot a tighter fit. This is known as wadding, and it was used to identify a murder weapon in one of the first ever applications of forensic science.

In 1784, the murder of a man took place in England. The victim's name was Edward Culshaw and he had been shot in the head. Suspicion fell on his friend, John Toms. The police found in Toms's possession a muzzle-loading pistol. In the wound in the dead man's skull the projectile was found, together with a crude wad. The wad had been torn from a newspaper, known then as a broadsheet. In Toms's possession, a broadsheet was found with a torn fragment missing. It matched the wadding in the wound exactly.

Efforts to improve the means of ignition went on. A fuse gave way to a mechanised means of bringing the burning fuse into contact with the main charge, but this was relatively short-lived and soon gave way, in turn, to the flintlock. A piece of flint was held in the jaws of the cock (so called for its supposed resemblance to a rooster) and, when released at the end of the s-shaped hammer, it would strike a piece of metal called the frizzen. The sparks generated would ignite the powder in the pan immediately below, which would spread to the main charge. I have included some photographs of a flintlock being discharged. There is a distinct delay between the ignition of the charge in the pan and the ignition of the main charge.

All the methods of discharge, however, still suffered from the same disadvantage: the gunpowder was very susceptible to getting wet, which made it quite useless.

The next great step forward was the invention of the percussion cap, which enabled muzzle-loading firearms to fire reliably in any weather. The percussion cap was a small container holding an explosive mixture that would explode if it was struck with a sharp blow. The cap was placed over a metal 'nipple' where the touch hole used to be. The resultant flash was transmitted to the main charge and hey, presto.

Percussion caps were slow to catch on. Settlers and farmers in remote places could always pick up a bit of flint from the surrounding hills. When your percussion cap ran out, you would be left with a funny-shaped club. Indeed, flintlocks stayed in fashion for a long time; many animals were hunted and many wars were fought with them.

The next great advancement was to combine the percussion cap with the powder in a single unit called the cartridge (see photograph). This more or less coincided with the development of breech-loading weapons. These weapons made use of a cartridge where the bullet, the powder and the ignition source were incorporated into a single self-contained unit. The need to keep the powder dry became far less demanding and the rate of fire improved. Development of breech-loading cartridge-firing weapons was swift, and one has today a bewildering array of rifles, handguns and automatic weapons of every calibre and type.

With the need for greater range and accuracy, the next step forward was the production of a barrel that had a set of spiral grooves, usually five or six, cut into its inside wall. Hence, we refer to these as rifled weapons. The rifling imparts spin to the projectile, which by this point had become bullet-shaped and so prevented tumbling, the gyroscopic conservation of angular momentum having a stabilising effect. Newer steels and high-power propellants resulted in high pressures developing in these modern weapons. The bullet was now driven down the barrel at an increased speed and under great pressure.

One of the most important functions of the forensic firearms examiner is to link the projectile or the cartridge case with the weapon that fired the shot. Because of the standards of proof required by the modern courts, this is a difficult task, but it was not always so. One of the first to walk this path was a Frenchman called Alexandre Lacassagne, who noticed that a bullet he had removed at autopsy had seven grooves in its surface. Having been shown several pistols, he found one belonging to a suspect which had seven grooves in its barrel. On the base of this slender evidence, a conviction was secured. This all happened in the spring of 1889. Unfortunately, it was to foreshadow much of what was to pass as forensic science. No one thought to ask how commonly a seven-grooved pistol was to be found.

Effectively what happened was that Lacassagne oversold the science. Although the attempt was the first step along the path to bullet identification, he had no notion that he was dealing with class characteristics only. There may have been many thousands of weapons with seven grooves. The issue has been discussed in the book *Criminal Investigation*. Class characteristics, as the name would suggest, are common to every item of that class that has been manufactured. Once idiosyncratic damage or wear is superimposed on the class characteristics, we can talk about individual characteristics, which are of use in characterising a particular weapon. This is discussed in more detail in the next section.

Interestingly, Lacassagne never took the subject any further.

The 'science' of bullet identification

The basic foundation or paradigm on which bullet and firearm identification is based may be stated as follows:

During the manufacturing process and during the lifetime use of a weapon, minute imperfections in the barrel, on the breech face and on the firing pin, as well as on all the other surfaces, including the extractor and the ejector, will be found. At the high pressures of the firing process, the soft brass of the cartridge case is forced up against these components and the imperfections become embedded or engraved in the brass. While the bullet travels down the barrel, it, too, picks up the minute imperfections on its surface. The marks imparted are said to be truly unique; no other weapon will show the same set of marks.

If these basic premises are wrong, then the whole subject of ballistics, tool marks, and firearm and bullet identification is called into question.

The investigation process starts with the recovery of a bullet either from a body or from the scene of a crime. If and when a suspect appears on the radar, his weapons are seized and are test-fired into a water tank. This allows a bullet to be recovered without being damaged in any way; water slows down the bullet remarkably efficiently. The recovered bullet, together with the crime-scene bullet, is examined using a comparison microscope. This piece of equipment is simply two ordinary microscopes joined by a comparison bridge, which allows the image from both microscopes to be viewed simultaneously through the eyepiece. Any marks and identifying features can be viewed alongside each other and compared.

The photograph labelled '[Ballistics comparison image 1]' in the photograph section shows a typical example of a 'match' between two test bullets known to have been fired from the same weapon. If one examines such a match, there are remarkable similarities on either side of the dividing line. There are also certain differences. If one looks at the bottom of the photograph, the lines across the dividing line are slightly out of register. There are also lines that differ on the one side from those on the other side, but all in all this is a fine example of correspondence of micro-striations on two bullets from the same revolver.

Turning now to the photograph labelled '[Ballistics comparison image 2]': this is an actual 'match' from a relevant court case marked on the court chart. The number and clarity of the points that 'constitute a match' in this case are not at all clear. In any reading of this so-called match, the subjective assessment is obvious. The photograph labelled '[Ballistics comparison image 3]' is another example of a 'match' being called on a paucity of corresponding lines. Seven lines are noted, when in reality only two or three are unequivocal.

Earlier researchers, such as David Q. Burd and Paul L. Kirk, have noted that 20 to 25 per cent of marks from different tools can coincide. J.L. Booker, in a 1980 article, states, 'the assumption that all striae are random is not one which can be freely made in the examination of bullets'. He goes on to explain that 'most tools used to manufacture gun barrels are prepared by grinding with abrasive wheels which resemble closely packed spheres of uniform diameter. (*Microscopically*, of course.) The spacing and depth of the primary marks on the tools are proportional to the size and fineness of the abrasive' (emphasis added). This means that the marks made by such tools are not randomly generated.

Booker shows a mark made by a random grinding with a 400-mesh abrasive, which produces an impressive number of coinciding marks when compared to another grinding on the same wheel. What is even more disturbing is that, when these randomly ground pieces of metal were shown to five experienced tool-mark examiners, all five reported finding seven or more consequently matching striations. This finding calls into question the pronouncements of Burd and Kirk, who claimed 'irrefutable identification on the basis of only five matching striae'. Walter Rowe, writing in *Forensic Science Handbook, Volume II*, notes that 'ample evidence exists that patterns of striations are not generated randomly'. He says that matches should be based on groups of consecutive striations rather than on individual striations. I am in agreement with this, except to say that I have rarely seen it practised. The second point he makes is 'that each firearm examiner has to establish his or her own criteria for what constitutes a positive match'. This conclusion, I am afraid, is not science. It is pure subjectivity gone wild.

So far, we have been discussing the marks impressed on the bullet by its passage down the barrel of the weapon. These are scrape marks. Another set of marks is used in the forensic identification of firearms, and it was described briefly above. When the bullet is propelled down the barrel, the brass cartridge case is forced back against the breech face of the weapon, where it may have the various imperfections of the breech face stamped into its soft metal. The impact of the firing pin in the primer also leaves an impression. In semi-automated pistols and rifles, the cartridge case is extracted from the breech after firing by a hook-like component called the extractor, and is ejected from the breech by the ejector. All of these actions mark the brass, and all of these marks have been used to identify and individualise weapons. If the marks on the breech face are not truly random, then individualisation of a weapon must take into account the non-random elements and discard them as having any probative value.

The moment it is established that the generation of the marks or striations on the bullet are not random, the most significant strut on which bullet identification rests is fatally weakened.

In Chapter 3 the differences between class characteristics and individual characteristics were discussed in relation to tyre or shoe prints, and it was established that the randomness of individual characteristics is what imparts value in the identification of a particular print. The same is true generally for firearms. If a genuinely random type or family of marks is imparted on the bullet on its way down the barrel, then these may conceivably be used to identify the weapon that made them. The marks must be random; they cannot be the common result of a class characteristic.

In the *Journal of the Forensic Science Society*, Booker explains in some detail why micro-striations on a bullet are not produced by a random process. He says, 'The assumption that all striae are random is not one that can be made freely in the examination of bullets.' That this fact seems to be common cause in the world of firearms identification may be gleaned from Rowe's article quoted above.

As mentioned, the micro-striations on the bullet are not the only way in which evidence left at the scene of a shooting may be linked to a given weapon. Classically, firearms identification 'experts' have used the marks imparted to the cartridge case by the breech face, the firing pin, the extractor or the ejector. Are these sufficiently individual to be used in this way? Tsuneo Uchiyama, writing for the Association of Firearm and Tool Mark Examiners (AFTE), has addressed this question. He makes the point that 'the discipline of firearm and toolmark identification is based on two empirical hypotheses. The first hypothesis is the consistency/reproducibility of markings which originates [*sic*] from the same firearm or tool. The second hypothesis is the existence of differences between markings originating from two different firearms or tools. *However, in reality, markings will change rather rapidly in the process of successive firing. On the other hand, markings from two different tools or firearms of the same make and type are sometimes similar because of high quality control during the production process*' (emphasis added). His conclusion is worth noting in full:

> Markings on one hundred successively fired bullets and expended cartridge cases were examined. Markings among different manufacturers of ammunition differed significantly even between consecutively fired bullets. Identification between bullets from different manufacturers was generally different. Diameter, weight and/or velocity of bullets will affect the reproducibility of striations on landmarks. A smaller number of striations were observed on the small diameter bullets. There were prominent parallel breechface markings on the primers of expended cartridge cases. The size of the primers' area of contact with the breechface and firing pin was different among each brand of ammunition but they were consistent within the same make of ammunition. There were concentric circular lines on hemispherical firing pin identifications. These circular lines were not exactly circular. The

> similarity of these circular lines between the same make of cartridge was rather high. On the other hand, the similarity of these circles between different makes of cartridge was low even between successively fired cartridge cases. The reproducibility of the diameter of these lines was rather high. However, the detail in these circular lines changed from firing to firing and therefore could not be used as individual characteristics for identification purposes.

Much of this is not new. Writing in 1959, Al Biasotti noted '15–20% matching striations on land or groove impressions between similar types of bullets known to be fired from different firearms of the same manufacture and type'. Of equal concern is the fact that 'the percentage of matching striations in bullets known to be fired from the same firearm was 21–38%', and of some importance is the weight given to a small number of matching striations. Biasotti and John Murdock, authors of 'The Scientific Basis of Firearms and Toolmark Identification', note that the weight given to these matching striations is subjective.

Brian J. Heard observes that 'there is no dispute, however, that out of the thousands of lines present in any one comparison, a number must by pure chance alone, show agreement … *This identification of a stria pattern can only be obtained through extensive experience in the matching of stria and cannot be taught in a book*' (emphasis added). This approach, I am afraid, is inimical to the scientific method. It reeks of subjectivity and is more appropriately applied to assessing an artwork or a piece of music.

To add to the problems faced by the firearms identification technician, tool marks left by the manufacturing process are a further source of markings on bullet cases. These produce inconsistent striations in ammunition known to have been discharged from the same firearm. Heard takes note of the problems inherent in manufacturing marks on the ammunition, giving an example of how confusing manufacturing marks can be. The example came to light during a laboratory accreditation exercise:

> A number of cartridge cases were submitted and the examiners were asked to determine how many weapons had been used. The problem appeared to be very simple and everyone returned the same results, four cartridges in one gun and two in another. The examiners had been rather unfair and had obtained two batches of the same make of ammunition, one of which had very pronounced manufacturing marks and the other none. Four cartridge cases had been fired from one batch and two from the other and, as the breechface marks were extremely faint and the firing pin featureless, the mistake was easy to make. The test was eventually withdrawn as every participating laboratory returned the same results.

Heard then enters the arena of statistics and, after some elementary probability theory, calculates the probability of two grooves from different weapons matching as 1 in 184 756. He goes on to calculate the total probability of a random match as 1 in 52 860 000 000. Sadly, this sort of statistical approach is flawed. Given the random matches already alluded to and the non-random generation of the striations, it is simply junk science.

In 2008, the National Research Council of the National Academy of Science formed a research committee to assess the feasibility, accuracy and technical capability of a national ballistics database. They found that 'the validity of the fundamental assumptions of uniqueness and reproducibility of firearms-related toolmarks has not been fully demonstrated'. They went on to say that 'crucial to the theory of firearms identification is that random imperfections created in the machining and filing processes are said to make the surface (and the negative impression of the said surface, left on the fired cartridge casings) unique'. In a crucial passage, they comment, 'firearms identification ultimately comes down to a *subjective assessment*' (emphasis added). This makes nonsense of Heard's probabilistic calculations.

It is worth looking at AFTE's basic requirement for calling a 'match'. 'Agreement is significant when it exceeds the best agreement demonstrated between toolmarks known to have been produced by different tools and is consistent with the agreement demonstrated by toolmarks known to have been produced by the same tool.' They fail, however, to state what the best match between marks made by different tools is in reality, so this again comes down to a subjective assessment by the individual.

We get closer to the nub of the problem when we consider what Rowe had to say in his essay 'Statistics in Forensic Ballistics': 'Given that there [is] a vast number of firearms currently available, rifled by at least five different methods, extrapolation of their results to all firearms is a *leap of faith*' (emphasis added). Rowe was referring to a study by G.Y. Gardner on bullets fired from four 0.38 Special Smith & Wesson revolvers. He goes on to say that 'there could be thousands of weapons rifled with each of the four rifling tools whose barrels contain identical or very similar imperfections and produce similar markings on the bullets fired through them'. It is vital to realise that the AFTE theory of identification is 'rooted in the recognition that "the interpretation of individualisation/identification is subjective in nature"'. The terminology used is couched in terms such as 'quasi-quantitative benchmarks', 'sufficient agreement', 'significance', 'likelihood', 'so remote', 'agreement in both quantity and quality', yet no specific empirical definition is given for these terms.

D.A. Stoney is quoted in Rowe's article, referring specifically to fingerprints: 'When more and more corresponding features are found between two patterns, scientists and layperson alike become subjectively certain that the patterns could not possibly be duplicated by chance.' This, he continues, happens in fingerprinting as in other branches of forensic science, '*[w]ithout any statistical foundation*' (emphasis added).

The conclusion drawn in the report 'Ballistics Imaging' is as follows: 'The validity of the fundamentals of uniqueness and reproducibility of firearms-related toolmarks has not yet been fully demonstrated.' Additional general research on the uniqueness and reproducibility of firearms-related tool marks would have to be done if the basic premises of firearms identification are not to imply the presence of a firm statistical basis when none has been demonstrated.

Firearms examiners tend to cast their assessments in bold absolutes, commonly asserting that a match can be made to the exclusion of all other firearms in the world. Such comments cloak an inherently subjective assessment of a match with an extreme probability statement of a match that has no firm grounding in science and unrealistically implies an error rate of zero.

Returning to the ballistics comparison images discussed earlier in the chapter, several things should stand out clearly. The first is the extreme variation in the quality of the marks deemed by a police ballistics 'expert' to be a match. The second is the extraordinary coincidence of 'matching' striations in two bullets fired from different weapons. A careful examination of the photograph labelled '[Ballistics comparison image 4]' will reveal that one set of marks was made by a land (raised portion), in the barrel, and the other was made by a groove. One weapon was a standard-issue Beretta 9mm pistol and the other weapon was a Tanfoglio 9mm Parabellum. Another interesting feature of this photograph is that, if I ask a person or a group of people to tell me how many lines seem to match, the answer is highly dependent on the prior information given to the person being asked. If the prior information is that this is truly a match, the answer is usually in excess of ten matching lines. If the information given is that this is not a match, then the number of lines corresponding falls to four or five. If no information is given, the number of matching lines is around seven.

These views are reiterated in the National Academy of Sciences report discussed in Chapter 10, *Strengthening Forensic Science in the*

United States. In tool-mark comparisons we have a most unhealthy situation. Heard puts it like this: 'The basic difference between an expert and a layman is that due to his experience and training the expert can observe and understand the features and phenomena which the layman would overlook.' This is unfortunately not so in that, in the courtroom setting, the judge, who is presumably a layman with respect to ballistics, should be able to see everything that is seen by the expert. Whatever the expert can see should also be observable by the layman once it has been explained to him.

Adina Schwartz, writing in *The Champion* of October 2008, has raised many of the troubling points alluded to in the preceding paragraphs. She notes the fundamental flaws in the basic premises of uniqueness and persistence of tool marks, commenting on the subjectivity of the 'match' and the issues of sub-class characteristics and their role in muddying the waters. We are left with identification as 'an art limited by the intuitive ability of individual practitioners'. This is neither healthy nor is it science.

In the *UCLA Law Review,* Jennifer Mnookin and her colleagues make the point that 'several of the most significant journals focussed on publishing pattern identification research simply do not comport with broader norms of access, dissemination or peer review typically associated with the scientific publishing'. The *AFTE Journal*, for example, is not freely available and requires written testimony from existing AFTE members. It has extremely limited dissemination beyond the members of AFTE – it is found only in eighteen libraries of the 72 000 libraries listed in World CAT, the largest catalogue of library materials – and completely lacks integration with any of the voluminous networks for the production and exchange of scientific research information. The journal engages in peer review that is neither blind nor draws on an extensive network of researchers. This is not an attitude in keeping with the openness that is a part of any true scientific research culture. And this closed-shop approach is not limited to the *AFTE Journal*: I recently encountered a similar approach in accessing the Forensic Science Society and their journal, *Journal of the Forensic Science Society*. In my view this limits debate as well as challenges to the status quo.

The meaning of all this for the man in the street is quite simple. Forensic identification evidence is often central to the outcome of a trial. The certainty of an identification is diluted enormously by the uncertainties that I have highlighted. The forensic science establishment is well aware of the problems, yet it persists in attaching improbable statistics to an identification that is highly subjective and whose fundamental premises are by no means proven; they may not even be valid.

The number of scientifically qualified researchers in the field is low. This means that witnesses who are scientifically illiterate are attaching an undeserved significance to evidence that may be dispositive in some cases.

The only hope is that the courts will act as a gatekeeper and keep the junk science out. This, however, is a forlorn hope when the type of evidence is 'I can't tell you what a match is but I know one when I see one.' When this type of bogus science is allowed to pass muster by the court, the unscrupulous prosecutor and, importantly, the uninformed defence, justice is not well served.

It is appropriate to end this chapter with a quote from mathematical physicist and engineer William Thomson (later Lord Kelvin): 'When you can measure what you are speaking about and express it in numbers, you know something about it; but when you cannot measure it, when you cannot express it in numbers, your knowledge is of a meagre and unsatisfactory kind; it may be the beginning of knowledge, but you have scarcely in your thoughts advanced to the stage of science.'

Chapter 16
GUNS FOR HIRE?

'The employment of private experts is open to serious objection, for they are only human, and it is natural for a man to sympathize with the side of the case from which his fee comes.'

– Harry Söderman

'The trouble with expert testimony is that it is setting the jury to decide where doctors disagree – but how can the jury judge between two statements founded on an experience admittedly foreign in kind to their own. It is just because they are incompetent for such a task that the expert is necessary at all.'

– Judge Learned Hand

'That which can be asserted without evidence, can be dismissed without evidence.'

– Christopher Hitchens

When experts take sides in a matter, the result is often an abandonment of objectivity. The nature of forensic science is that it enables inferences to be drawn from scientific principles applied to the residual evidence at the scene of the crime. The inferences are critical depending on what facts are presented to the expert. Minor variations in the starting-point 'facts' can lead to diametrically opposed viewpoints. Coincident with my starting to write this chapter, reports appeared in the local newspaper regarding the high-profile trial of a woman who stands accused of murdering her husband, a judge of the High Court. Apart from the almost total lack of proper investigation at the scene of the death, the case has descended into a duel between the pathologist for the state and the pathologist for the defence. The state expert, Dr Sipho Mfolozi, ruled out natural causes. Professor Gert Saayman, for the defence, disagreed, saying that in his opinion 'natural causes should be the primary consideration in determining the cause of death'. It is clear that the state is of the opinion that the death of this judge was a result of a deliberate act on the part of his wife or someone acting on her instruction.

Saayman and Mfolozi evidently hold divergent views. The only two things that can be said about this is that one of these two pathologists is wrong, and this kind of duel brings the profession into disrepute. Mfolozi's statement that he can 'rule out natural causes' looks very much like the negative corpus approach that I so disparaged in the chapter on fire investigation. These two experts should have met before the trial to discuss their points of agreement and their points of divergence. This kind of diametrically opposed testimony is not uncommon in the courts, not only in South Africa but in countries across the world. This whole problem has been given voice in *Forensics Under Fire* by Jim Fisher, mentioned in Chapter 10.

So far in this book, expert testimony and the discipline of forensic science have been discussed. The aims of any honourable justice system are to protect society by weeding out the wrongdoers and making sure that they receive some form of official sanction that is commensurate with the crime and has a deterrent effect both on the wrongdoer and on others wishing to emulate him or her. In days of yore, the stealing of a loaf of bread by a child was a hanging offence. The severity of the punishment stood in such sharp contrast to the triviality of the crime that, thankfully, we no longer apply such barbaric law. The key to any justice system is that the guilty are the ones who need to be punished. If we punish the innocent, then the law fails in its primary task and will, deservedly, fall into disrepute.

The vastness of the disparity between the state's power and resources and those of the individual make it imperative that safeguards, checks and balances are built into the system to prevent injury to the innocent. Unfortunately, as demonstrated in the preceding chapters, the structure of justice departments and police forces worldwide make this notion of justice an almost unattainable goal. Daily we witness the spectacle of people in high office escaping judicial scrutiny and retribution because they are able to manipulate the levers of office to deflect any scrutiny of their actions. In South Africa, the ruling party under Jacob Zuma has routinely manipulated the National Prosecuting Authority to deflect the criminal trial of Zuma for fraud. The ruling party has saturated the bench with minions and political cronies often of mediocre talent in order to achieve its political will. The appointment of the current Chief Justice may be said to fit that description – there were far more able candidates who were overlooked. There are, however, other areas where justice is subverted by the state, and it is with regard to these areas that explanation and exposure are long overdue.

In all political and judicial systems that I am aware of, the forensic science services are part and parcel of the police departments. The theory is that police will conduct an intelligent inquiry and, on the basis of properly collected evidence, will identify a potential suspect or suspects. The forensic scientists and state pathologists will impartially and rigorously examine the evidence by trusted, accepted and infallible scientific procedures.

Unfortunately, the ideal falls far short of reality – so much so that in 2009 the National Academy of Sciences in the US was forced to write *Strengthening Forensic Science in the United States*. As stated earlier, they have written that, in some cases, substantive information and testimony based on faulty forensic science analysis have led to unjust convictions of innocent people. We know that faulty forensic information derives from several sources: collusion with the prosecutor to edit out evidence that would have an exculpatory value to the defence, incompetence in analysing and interpreting the scientific evidence, overstatement of the confidence value of the evidence, and plain dishonesty and fraud.

Let us start by recalling the dishonest behaviour of the forensic experts in *State* v. *Van der Vyver*. The case was characterised not only by botched police work and faulty forensic analysis, but also by blatant lies and attempts to deceive the court on the part of expert witnesses (see Chapter 6).

Are we alone in this deplorable state of affairs? The answer is no. We have already encountered Englishman Alan Clift and his willingness to alter his forensic report to comply with the prosecutor's request.

During the Van der Vyver trial, the prosecutor, Carien Theunissen, approached a police expert, Charlene Otto, and put her under some pressure to change her findings in the report she had given the prosecutor previously. Otto would have none of it and sent the prosecutor packing. That is how forensic science should work, and how its practitioners should behave. It is not, however, how prosecutors should behave.

A brief trawl on the internet will expose some of the more egregious cases of forensic fraud. The cases involve personnel ranging from junior members of laboratory staff right the way up to the top lab directors in some of the most prestigious institutions in the world. Forensic fraud has no boundaries. I shall list some of the cases in alphabetical order:

Talibah Akili, foster-care coordinator at the Ibero-American Action League Minority Foster Boarding Home Program: convicted of fraud and perjury when she lied under oath about her qualifications. She had no tertiary education despite claiming to have multiple post-graduate college degrees and to be a certified social worker, forensic counsellor and substance-abuse counsellor.

Sandra Anderson, dog handler: planted evidence for her dog to find.

Concepcion Bacasnot, forensic chemist: gave false testimony about blood-type evidence for the Baltimore Police Laboratory.

Dr John Brown, DNA analyst: admitted to lying under oath in a case in Seattle in 2000.

Tom Callaghan, FBI agent: urged crime-lab officials not to perform certain searches that would reveal error rates in the fingerprint database.

Timothy Dixon, Illinois State Police Crime Laboratory: falsely exaggerated statistics in a rape case to strengthen the prosecution's case.

Pamela Fish, biochemistry section chief in Chicago: gave false testimony in nine cases.

John Fitzpatrick, Orlando Crime Laboratory: falsified data in a test that was checking the quality of work in the lab.

Mary Furlong, Illinois State Police Crime Laboratory: failed to report that the semen found in the deceased's vagina was different from that of the accused.

David Harding, New York State Paratroopers Forensic Unit: admitted to planting fingerprint evidence to secure convictions.

Wallace Higgins, FBI supervisor: altered reports of another FBI operative, Fred Whitehurst, without his permission.

Sam A. Kaminsky, Garden City Police Department, Georgia: falsified fingerprint evidence by relabelling inked prints taken by the police as crime-scene prints – of course they matched!

Allison Lancaster, San Francisco Police Crime Laboratory: faked lab results in drug cases.

Brian W. Meehan, DNA 'expert': colluded with District Attorney Mike Nifong to withhold DNA results favourable to the defence in the Duke Lacrosse trial.

Gene Morrison, Yorkshire Police: bought his qualifications from a phony university on the internet.

Louise Robbins, footwear and footprint examiner: claimed to be able to identify perpetrators from footprints and was debunked by the world forensic community for testifying to the impossible. (More on Louise later.)

Larry F. Steward, director and chief forensic scientist, US Secret Service Crime Laboratory: indicted for perjured testimony regarding ink analysis in the Martha Stewart case.

Fred Zain, serologist: gave perjured evidence in scores of cases relating to blood grouping in West Virginia. (More on Fred later.)

This is by no means a complete list of fraudsters in the world of forensic science. The information is readily available on the internet simply by doing a search for the words 'forensic fraud'.

In *Steeped in Blood* I wrote about the Mandrax case, a South African example of forensic fraud. The analyst was too lazy to prepare seven samples from the 'drugs' that had been seized, so he simply analysed the laboratory reference sample seven times. Similarly, in the police shooting discussed in Chapter 3, the evidence of the blood-spatter expert, Colonel Kock, was invalid: he conceded that he had no knowledge of anatomy or physiology and was basing his evidence solely on the blood-spatter analysis.

Brandon L. Garrett and Peter J. Neufeld have written about invalid forensic testimony and wrongful convictions in the *Virginia Law Review*. Theirs was one of the first studies of forensic evidence that had helped send innocent people to prison. It was conceived against the backdrop of the exoneration and release of many of these prisoners, some from death row, often as a result of exculpatory evidence coming to light in the form of DNA analyses, sometimes years after the trial and incarceration. The authors examined the trial records of 137 of the 156 persons who had been exonerated. The type of forensic evidence that assisted in convicting them included blood-group identification and hair, bite-mark, shoe-print, soil, fibre and fingerprint comparisons. A review of the trial evidence showed that experts misstated the empirical data or gave evidence that was wholly unsupported by empirical data. And it was not just a few rogue analysts; it occurred right across America, despite well-developed principles of law on when and where scientific evidence should be allowed. Judges were slow to grant relief and to perform their stated role as gatekeepers. Of equal concern was Garrett and Neufeld's finding that defence attorneys rarely cross-examined witnesses on scientific matters. In my time as a practising forensic scientist, I have found that lawyers have an aversion to entering the scientific field to cross-examine experts. They prefer to start with issues such as chain of evidence and amount of experience of the expert. The whole scientific basis for the expert's views is seldom challenged.

The advent of DNA-based forensic science has brought about significant changes in the way forensic science is viewed. Many of the so-called staple forensic science disciplines have been shown to have little or no hard scientific basis and, in many instances, to have played a significant role in procuring wrongful convictions.

Gary Dotson, serving a twenty-five-to-fifty-year sentence for rape, was released when DNA testing of the semen sample from the accuser exonerated him. At this point, Dotson had been refused a pardon by the Governor of Illinois despite the complainant having recanted her initial statement. In her recantation, she said that she had fabricated her original statement to conceal from her parents consensual intercourse with her boyfriend at the time. Again we have in this matter deliberately misleading evidence given by the state blood-group analyst, who omitted to tell the trial court that the accused shared his blood group with the accuser, and thus there was no way of identifying his group in her vaginal sample. This same deliberate error was made in the Alan Clift affair.

The Innocence Project, a litigation and public-policy organisation dedicated to exonerating wrongfully convicted people, found that 57 per cent of cases involving serology (the analysis of blood groups) featured invalid forensic science testimony. This is not an insignificant number. What is more concerning is that only DNA could play a role in exonerating the accused in these cases. There is nothing to suggest that this error rate does not extend into other types of trial, where the spotlight of DNA analysis cannot reach.

In the case of Ray Krone, who achieved great fame as the 'Snaggle Tooth Killer', a bite mark was central to the matter. Krone had a skew tooth, and the 'forensic odontologist' who testified for the prosecution stated empirically, 'That tooth caused that injury.' Reference to *Strengthening Forensic Science in the United States* quickly shows that 'forensic disciplines involving impression evidence such as bite mark and shoe print comparisons have not developed any objective criteria at all by which to judge assertions about the likelihood that crime-scene evidence came from a particular defendant. Nor do any empirical data exist to support a claim that a bite mark is uniquely identifiable as belonging to a particular person.'

Standing on my bookshelf is a work by an associate professor of anthropology at the University of North Carolina, Dr Louise Robbins, titled *Footprints: Collection, Analysis and Interpretation*. If Robbins had not strayed from the straight and narrow, she would not be as universally vilified as she is. She was one of those witnesses who could see things that others simply could not see at all. Descriptions of her work range from 'complete hogwash' and 'cockamamie stuff' to 'pseudo science'. William Bodziak said of her, 'Nobody else has ever dreamed of saying the kind of things she said.' Robbins is unfortunately typical of the kind of witness who uses junk science to convince a court of a prosecution-based theory. Often the rationale rests on an 'experience' that is difficult to test in court, and on an ability to see what no one else can.

We have our own home-grown witnesses like Robbins in South Africa, such as the forensic expert referred to in the context of insurance-related matters who not only falsified his qualifications in a number of cases but also claimed to have written a book which he never had. In a case involving an arson investigation done by this man, he claimed certain evidential signs in his expert report. When I pointed out to him in the experts' meeting that these vital signs were not shown in the photographs, he brazenly said that he 'was there and saw them', despite not being able to demonstrate them photographically. This expert recently investigated a fire in Cape Town, where he unequivocally stated that there were four separate origins of the fire. An investigation on my part revealed the error, and for once the insurance company sent a second investigator, who agreed with me. The claim was paid.

Are these merely rotten apples in the barrel? I think that this is too simplistic an approach to take. Removing the rotten apples is merely a symptomatic treatment. One needs to investigate why there are so many rotten apples and treat the root cause.

At one time, the head of forensics in South Africa was a particularly repulsive individual called Lothar Paul Neethling (an encounter of mine with him was described in Chapter 3). He was of German descent, having been brought out to South Africa by the Nationalist government. One case involving Neethling needs to be told for two reasons. Firstly, it gives insight into a truly evil man and, secondly, it demonstrates the way in which that kind of wickedness filters down the ranks and into the laboratory, casting a dark shadow over all of the laboratory's work. The background to this case has a long history, starting with the murder of Steve Biko while in police custody. This had resulted in an inquest, where the police and district surgeons were savaged in the witness box by the incisive cross-examination of Sydney Kentridge, who acted for the family. The case attracted worldwide publicity and generated massive embarrassment for the government and the police at the time.

Neethling saw himself as the man to solve the thorny problem of adversaries of the apartheid regime without leaving obvious clues behind. Beating the victim to a pulp became unfashionable; a more elegant way needed to be developed. In the pursuit of this goal, Neethling, who was an organic chemist by training, set about developing a drug that was hard to detect and that would precipitate a heart attack in the victim. This drug was tested on sheep in the Pretoria police laboratories.

Clearly the potion worked very well on sheep, as Neethling summoned the top police operative who had started the police assassination group at Vlakplaas. This top cop was Dirk Coetzee, whose name has become infamous in South African history. There was a dissident Coetzee needed to get rid of, Sizwe Kondile. He travelled to Neethling's office in Pretoria, where he was made to sit and wait while Neethling fetched the 'deadly potion' from the safe. Neethling gave the poison to Coetzee with instructions on how to use it and orders to record the effects of the drugs on the victim and to report back. The potion turned out to be unsuccessful. Eventually the police officer involved, Koos Vermeulen, shot the dissident in question. His body was cremated on the banks of the Komati River near the Mozambique border.

All of this would have stayed secret but for the activities of another Vlakplaas member, Butana Almond Nofomela. He had been trained as a killer by the police at Vlakplaas and, just so long as he limited his killing to opponents of the apartheid regime, he was fine. However, one day he decided to do a little business on his own account, and he murdered a farmer and his wife in the sleepy town of Brits, about forty minutes south-west of Pretoria. He was duly caught, tried and sentenced to death by hanging. While sitting on death row in Pretoria, his spirits were lifted now and again by his former superior officers popping in to let him know that they were pulling whatever strings they could on his behalf. On 12 October 1989, Nofomela received the notice of set-down for his execution, and five days later he was visited by two members of the security police who, using a delicate turn of phrase, informed him that there was nothing more they could do for him; he would have to 'take the pain'. At this point, he felt betrayed by his superior officers and decided to reveal everything he knew about Vlakplaas, Dirk Coetzee and all the dirty tricks of the SAP. I have included his death-cell affidavit in the appendices, as it makes for interesting reading and is a small piece of our darker history that should not be forgotten. The affidavit begins: 'I am a thirty-two-year-old male, presently under sentence of death. My execution is scheduled for tomorrow morning, 20 October 1989 at 07:00.'

He managed to contact the organisation Lawyers for Human Rights, which brought an urgent application to the High Court, and the execution was stayed. Nofomela gave his evidence. Coetzee, realising that he would be implicated by this evidence, made contact with Jacques Pauw and Max du Preez, who were running the boutique newspaper *Vrye Weekblad*. Pauw and Coetzee secretly travelled to Mauritius, and there Coetzee unfolded the incredible story of the SAP death squads. Most importantly, his testimony implicated Lothar Neethling in the evil business of using and abusing the state-run forensic laboratory for illegal and supremely wicked activities. Pauw writes about this in *In the Heart of the Whore: The Story of Apartheid's Death Squads*.

The purpose of telling this story is to illustrate how government-sponsored and run facilities can go rotten from the top all the way down to the bottom. It would have been impossible for the personnel of the laboratory not to have known about the goings-on. They did nothing. I am a believer that silence is equivalent to complicity.

Government agencies, despite having the best of funding and an abundance of specialist posts and personnel, do not have a monopoly on honesty, competence or integrity. The disastrous mistakes made by the top fingerprint experts at the FBI do not inspire confidence in the work done at less prestigious police departments. The sheer incompetence of the Scottish fingerprint experts in the case of Shirley McKie is another example.

The manipulation of results and data by forensic scientists who should have no intent in the outcome of the case is clearly widespread. Overselling the value of the evidence by the state experts is another area I encounter in my own practice on a regular basis. The need for the defence to challenge and check every iota of state forensic evidence has never been so acutely necessary. It must be said that the misuse of forensic science in court has as much to do with unscrupulous and dishonest court officials as it has to do with the forensic scientists who abuse their knowledge for a dishonourable purpose.

As noted in Chapter 10, the television version of forensic science is as far removed from reality as it is possible to get. The real scene is peopled by forensic scientists, some of whom are beyond reproach and others who are not. These scientists have given evidence that is questionable, erroneous and, in some cases, like many of the matters discussed above, downright fraudulent. In 'Beyond Bad Apples: Analyzing the Role of Forensic Science in Wrongful Convictions', William C. Thompson describes the 'rotten apples' as coming in 'three flavours': careless, incompetent and malicious. Sometimes the sub-standard work can permeate entire departments: the entire DNA/Serology Unit of the Houston Police Department Crime Laboratory, for example, was closed down owing to a whole range of incompetence, fraud and misrepresentations. This had been going on for over ten years before it was exposed. Actors who are expected to provide an independent layer of review may, due to imperatives arising from their occupational role, not only fail to expose problems but take steps to avoid having those problems exposed in subsequent levels of review.

The disastrous failure of forensic science to do its intended job is nowhere better illustrated than in Garrett and Neufeld's 'Invalid Forensic Science Testimony and Wrongful Convictions'. What is disturbing is their conclusion that 'the system has still not responded with a full investigation into these known miscarriages of justice or routinely conducted investigations to ensure accurate deposition'. Furthermore, as is evident from the cases discussed in this book, 'the adversary system cannot be depended upon as an adequate safeguard. The defence bar lacked the expertise and resources to detect and address invalid forensic science effectively in most of these cases and judges did not remedy most errors brought to their attention.'

These last points are discussed in the next chapter, which deals with cases in which the dishonesty of the prosecutors, coupled with the unscrupulous behaviour of state forensic scientists, has led to great miscarriages of justice. We shall also see how the scientific illiteracy of many of the defence lawyers allows this situation to prevail.

Chapter 17

FIAT JUSTITIA RUAT CAELUM: LET JUSTICE BE DONE THOUGH THE HEAVENS FALL

'Law and order exist for the purpose of establishing justice and when they fail in this purpose they become the dangerously structured dams that block the flow of social progress.'

– Martin Luther King, Jr

'The hope of a secure and livable world lies with disciplined nonconformists who are dedicated to justice, peace and brotherhood.'

– Martin Luther King, Jr

That, of course, is the theory – namely that justice must be done even if the mighty are laid low.

Lord Alfred Thompson Denning was at one time the most senior judge in Britain, who penned a book of legal meanderings, *What Next in the Law*. It so happened that the case of the Birmingham Six (see Chapter 11) came before the court while he was Master of the Rolls. It is now understood that these six men were convicted on the completely false forensic evidence of Frank Skuse, who used a forensic test that would have given positive results on any number of completely innocent and innocuous substances. Not only that, but it is now recognised that confessions were beaten out of the men by members of the Birmingham Criminal Investigations Division.

The six men brought charges against the West Midlands Police in 1977, which were rejected by Lord Denning, who wrote the following astounding paragraph in his judgment:

> Just consider the course of events if [the Six's] action were to proceed to trial ... If the six men failed it would mean that much time and money and worry would have been expended by many people to no good purpose. If they won, it would mean the police were guilty of perjury; that they were guilty of violence and threats; that the confessions were involuntary and improperly admitted into evidence and that the convictions were erroneous. That would mean that the home secretary would have either to recommend that they be pardoned or to remit the case to the court of Appeal. That was such an appalling vista that every sensible person would say, 'It cannot be right that these actions should go any further.' They should be struck out either on the ground that the men are stopped from challenging the decision of Mr Justice Bridge or alternatively that it is an abuse of the process of the court which it is, the actions should be stopped.

What Denning is saying is that, if the police were capable of beating the living daylights out of a suspect and obtaining a 'confession' and then presenting perjured evidence, to expose it is such an 'appalling vista' that he would rather let six innocent men rot in jail for a crime they did not commit. This paragraph is made more grotesque by Denning's writing in *What Next in the Law*:

> These courts have the authority – and I would add the duty – in a proper case, when called upon, to inquire into the exercising of a discretionary power by a minister or his department. If it is found that this power has been exercised improperly or mistakenly so as to infringe unjustly on the legitimate rights and interests of the subject, then those courts must so declare. They stand as ever between the executive and the subject, as Lord Atkin said in a famous passage, alert to see that any coercive action is justified in law. To which I would add alert to see that a discretionary power is not exceeded or misused.

I am not sure from what planet Denning was writing his judgment. The issues of police corruption, framing innocent people and serious other misbehaviour in England were well known. They formed the subject matter of a book by James Morton called *Bent Coppers: A Survey of Police Corruption*, in which the case of Chris Mullin is described.

On 13 December 1991, Chris Mullin, the Labour MP for Sunderland South, raised the matter of a two-year-old case in an adjournment debate in the House of Commons, saying, 'The West Midlands Police have been caught fabricating evidence.' He alleged that they did so in the knowledge that they could rely on their superiors, including the chief constable himself, to make sure that the truth was covered up. They could rely, too, on the Crown Prosecution Service to connive in the disappearance of inconvenient evidence.

This mindset rings out from Denning's judgment. He is willing to let injustice stand rather than expose the serious deficiencies in the system. It is so discordant with many of the pious views expressed in *What Next in the Law* that it is difficult to believe that the two pieces of writing were penned by the same man.

There are more examples of the unwillingness of judges to mess with government. Denning contributed more of the same in the 1960s Profumo inquiry. John Profumo was having an affair with Christine Keeler, who was simultaneously romantically involved with Eugene Ivanov (also known as Yevgeni Ivanov), a naval attaché at the Soviet embassy in London. Denning managed to incorporate deference for the government of Harold Macmillan with viciousness reserved for Stephen Ward, who was Keeler's friend and who had by that time committed suicide. If these kinds of double standards are applied by judges of the top echelons in the UK, how much more so could it happen in places where public vigilance is not as well honed as it is in Britain?

When Dirk Coetzee told his story about the dark side of the SAP and their role in crime and political assassinations, the matter resulted in a judicial inquiry under Judge Louis Harms, who has been in most ways an absolutely exemplary judge. His involvement in this commission into death squads, however, is a blot on his record. He allowed government secret agents to sneak into his court wearing ridiculous disguises. When Coetzee gave his evidence, Harms was so overcome by the 'appalling vista' opening up in front of him that in a rather unjudicial outburst he called Coetzee's evidence a lot of 'bullshit'. No matter where they are found, judges have often allowed injustice to prevail in cases where they were simply too disinclined to rock the governmental boat. As it happens, the unfolding of history in South Africa has revealed that Coetzee was right and Judge Harms was wrong on the issue of hit squads in this country.

Judges have a vital role to play in nurturing and protecting justice. The majority do, but there are those who see the system differently and who largely maintain the status quo. The system can be manipulated to ensure that the 'right' judge hears the case, especially in criminal matters. This happened frequently in South Africa during the apartheid years, when detainees who had been viciously and seriously assaulted by the Security Police would be brought before selected magistrates who would not see the often serious injuries inflicted on them.

In America, the landmark case of *Daubert* v. *Merrell Dow Phar-*

maceuticals Inc. made this gatekeeping role (as far as forensic science is concerned) very clear. As noted in Chapter 12, the purpose of the *Daubert* ruling is to keep junk science out of the courts. Following the Mayfield case (discussed in Chapter 6), challenges have been launched to confront the total admissibility of fingerprint evidence, but these challenges have not found much traction in the courts. If any case demonstrates the subjectivity of fingerprint evidence, it is the 2007 American case of *Maryland* v. *Rose*, in which the court excluded fingerprint evidence on the basis that 'it was a subjective, untested, unverifiable identification procedure that purports to be infallible'.

This decision did not last, as it was overturned by a higher court – a huge pity, because the initial judgment got it just right. These problems certainly do exist and the courts would do well to exhibit a little more scepticism in accepting this kind of evidence. The superior court in the case came to the conclusion that 'fingerprint identification based on the ACE-V method is generally accepted in the relevant scientific community'.

At this juncture it would be a good idea to examine the so-called ACE-V method, if it can be called a method. ACE-V is an acronym for Analysis, Comparison, Evaluation and Verification. It is proudly trotted out in fingerprinting, handwriting comparison, arson investigation and wherever else the forensic expert wishes to clothe unworthy science in a little false respectability. Jennifer Mnookin, in an important paper titled 'The Courts, the NAS and the Future of Forensic Science', writes cogently about ACE-V.

The first analytical step is to look closely at the item to see if it is of any value. The second is to document any detail that may be relevant to the identification. If there are enough details to compare, then a comparison is done between the same features in the reference items, be it a bullet, fingerprint, handwriting or anything else. The comparison is then evaluated to see if it constitutes a valid 'match'. It needs to be said up front that this is much ado about nothing. One could as easily motivate using this method to buy a second-hand car. One might look at the car carefully to see if it is worth buying. One might detect various features that may affect its desirability. One might compare it to other similar cars and, finally, one might ask a friend to assist in the evaluation of the decision. This does not make the process scientific.

The ACE-V method is often abused. For example, a local fire investigator will make use of ACE-V and the person who does the verification is his partner. They will sign reports verifying each other's findings. So much for an independent verification.

ACE-V's relationship to the scientific method is at best tenuous. It really amounts to nothing more than two examiners looking carefully at a set of evidence. To have any value at all, the person doing the verification must be truly independent and must be unaware of the first examiner's findings. This seldom happens. Normally the second person is an associate of the first, as in the above example, which hardly leads to an independent assessment. It is worth mentioning that the Scottish fingerprint officials in the McKie case all made use of ACE-V, but their close colleagues could not see what turned out to be obvious differences. The same was true in the Mayfield case. The questions that make any endeavour scientific are not present in ACE-V. There is no attempt to build on a firm data foundation because, in most of these pattern-comparison fields, none exist. Individual experience does not count as a database. ACE-V gives no evidence of error rates, and it takes no account of the scientific methodology to prevent various forms of bias creeping in. There is no basis for examining control data as opposed to the questioned thing. Ultimately the method depends on an unacceptable degree of subjective personal experience.

The reason that I include this discussion here is because judges are the guardians of what should be allowed into the courtroom as far as scientific evidence is concerned, and all too often they fail to implement this important function. It is a matter of fairly simple logic to see that merely following the steps of ACE-V does not ensure that the components are performed with the necessary scientific rigour or with the controls and safeguards that science would require as a starting point.

Judges are loath to get involved in the science. There are notable exceptions, but, as revealed in earlier chapters, most weigh the scientist or witness rather than the evidence. *Strengthening Forensic Science in the United States* points out the 'utterly ineffective' way in which judges have approached the thorny issue of evaluating and distinguishing good science from bad. The report suggests that 'it might be too late to effectively train most lawyers and judges once they enter their professional fields. Training programs are beneficial in the short term, because they offer responsible jurists a way to learn what they need to know. For the long term, however, the best way to get lawyers and judges up to speed is to offer better courses in forensic science.' Mnookin puts it like this: 'Forensic science experts should not continue to be given free rein to testify in the manner in which they have done up to now. Judges need to develop a variety of thoughtful approaches – a toolkit of sorts – with which they can assess admissibility and thus, the toolkit should absolutely include outright exclusion in some circumstances.' Another safeguard would be to have judges in scientific matters sit with scientifically trained assessors to assist them.

The sociologist Robert Merton King once said, 'Most institutions demand unqualified faith, but the institution of science makes scepticism a virtue.' Unfortunately, this spirit of scepticism is often lacking in the judicial approach to scientific evidence. This is particularly so when the challenge is made against state evidence.

In the US, there are well-known cases dealing with the admissibility of scientific evidence. This has metamorphosed into the *Daubert* trilogy and includes *Daubert* v. *Merrell Dow Pharmaceuticals*, *General Electric* v. *Joiner* and *Kumho Tire Co.* v. *Carmichael*. *Daubert* is possibly the better known of these. However, even when there are clearly defined guidelines, judges can get the interpretation wrong. In *United States* v. *Harvard*, the court became confused about the meaning of the testing of the evidence. In this instance, the judge relied on the fingerprint evidence because it 'had been tested in adversarial proceedings with the highest possible stakes, liberty and sometimes life'. This, of course, is a complete misapprehension of what testing as envisaged by *Daubert* is all about. The testing refers to scientific testing, not the results of legal adversarial combative testing in the hothouse environment of the court. A good example of why this is not the appropriate way to test expertise or, for that matter, the value of an expert's testimony, is provided by the famous 1931 case of *R* v. *A.A. Rouse*.

The case revolved around the 'murder' of an unknown man. It was alleged that Alfred Arthur Rouse had picked up the victim, murdered him and then incinerated the body in his car. Rouse advanced the defence that the fire was accidental. The case was prosecuted by Norman Birkett. Some time into the defence, an expert witness, Mr Arthur Isaacs, was called. He was an engineer with a good deal of experience in car and other vehicle fires. The prosecution alleged that Rouse had deliberately loosened the coupling on the fuel line to fuel the fire. Isaacs was of the view that the coupling had become loose during the fire and it was on this point that he was cross-examined by Birkett, in what has become an infamous passage:

What is the coefficient of the expansion of brass?

I beg your pardon?

Did you not catch the question?

I did not hear?

What is the coefficient of the expansion of brass?

I am afraid I cannot answer that question off-hand.

What is it? If you do not know, say so, what is the coefficient of the expansion of brass? What do I mean by that term?

You want to know, what is the expansion of the metal under heat?

I asked you: What is the coefficient of the expansion of brass? Do you know what that means?

Put that way, probably I do not.

You are an engineer?

I dare say I am.

Let me understand what you are. You are not a doctor?

No.

Not a crime investigator?

No.

Not an amateur detective?

No.

But an engineer?

Yes.

What is the coefficient of the expansion of brass? You do not know?

No; not put that way.

This cross-examination of Isaacs had the effect of disconcerting the man and demeaning his evidence in the eyes of the jury. Rouse was hanged and I have no doubt that this cross-examination helped substantially in tightening the noose. Scientifically, of course, this sort of cross-examination is nonsense. Firstly, there are many different brasses with significantly different compositions, ranging from gilding metal with a composition of 95 per cent copper and 5 per cent zinc to white brass, which contains 50 per cent zinc. In addition, some brasses contain other elements, such as lead and arsenic, which modify the properties of the alloy considerably. There are easily more than twenty different commonly found brasses.

From this, one can clearly see that the question posed by Birkett is a meaningless one designed only to produce confusion. A more experienced witness would have shown Birkett to have asked a foolish question. The question, and many others like it, are born in ignorance and nurtured in bluster and are scientifically worthless, although they may seem quite clever to an ignorant jurist. This is a good example of why the courts provide a poor test of science. In *Harvard* the judge clearly failed to understand the difference between scientific testing and adversarial testing in court. Again, Mnookin deals with the failure of courts to understand the nub of scientific evidence, quoting from *Harvard*: 'In some disputes despite the absence of a single quantifiable standard for measuring the sufficiency of any latent print for purposes of identification, the court is satisfied that latent print identification easily satisfies the standards of reliability in *Daubert* and *Kumho Tire*. In fact, after going through this analysis, the court believes that latent print identification is the very archetype of reliable expert testimony under those standards.' The court's view, Mnookin states, 'strains all credibility'. Only by putting one's head in the sand could one possibly conclude that the archetype of reliable evidence under *Daubert* is latent fingerprint evidence that has been tested adversarially but not scientifically, lacks meaningful error-rate information, and operates without statistical foundation or any validated objective criteria for determining a match. This indicates that the court in question understands neither *Daubert* nor the nature of the science involved.

In some instances, judges find themselves on a knife-edge. In another famous American case, *United States* v. *Llera Plaza II*, Judge Louis Pollack threw out fingerprint evidence in the first hearing. This produced a chorus of outraged howls from law-enforcement officialdom, especially the FBI. This first judgment found, quite correctly in my view, that fingerprinting technology did not pass *Daubert* muster. In a subsequent judgment reviewing the first judgment, Judge Pollack changed his mind. This followed the outcry by the government that prosecutorial effectiveness would be seriously compromised – the government urged the court to reconsider its order. Quite frankly, this is outrageous. Firstly, the government should not put pressure on a judge and, secondly, the judge should not submit to this pressure and reverse his order.

It has been my experience over some thirty years of practice in and around the courtroom that jurists fall into two broad categories:

those who will try to get to grips with and understand the scientific principles of the issue before the court, and those who understand nothing of the scientific evidence and rely on legal sleight of hand, like Birkett in *R* v. *A.A. Rouse*.

I remember a matter in the Western Cape area involving the causation of a fire that had swept through thousands of acres of government-owned forest. The fortnight before the trial, I consulted at length with the senior advocate who was running the defendant's case. In detail, I took this man (and his junior) through the evidence and explained the scientific basis for fire behaviour. It was apparent from the expert summary provided by the plaintiff in the matter that there were insurmountable problems with his testimony, which, if properly exposed in court under cross-examination, would have rendered his testimony completely worthless.

When the time came for the senior counsel to cross-examine this particular witness, his first few questions and his handling of the answers to them demonstrated that the advocate had understood not one word of my explanation and advice during the previous set of consultations. I caught his attention and the matter stood down while we went back to his chambers. I remember saying to him that his cross-examination was completely ineffective and that all my explanations the week before might as well have been in Spanish for all the understanding he displayed. Eventually the cross-examination was completed, the advocate reading from a script of questions that I had given him and taking frequent breaks to deal with the answers. Eventually we demolished the witness's evidence in its entirety, but this could have been done far more elegantly by counsel who understood the principles of the subject.

In South Africa, the case law relating to fingerprint evidence was decided in two cases. In *State* v. *Nala*, the court made the following statement: 'Where a trial court investigates the evidence of a fingerprint expert regarding points of identity it does so, not in order to satisfy itself that there are the requisite number of points of identity but so as to satisfy itself that the expert's opinion as to the identity of the disputed fingerprints may safely be relied upon.'

In a later case, *State* v. *Malini*, the court said, 'It is clear that the courts have taken the view that fingerprint identification is a matter for experts. Even though it cannot be clearly demonstrated to a court, the question is not so much whether the court can see the similarities or dissimilarities indicated by the expert but whether it can trust the expert and rely on his statement and opinion.' The judgment goes on to say, 'The attempt to examine meticulously the similarities and dissimilarities allegedly found by the experts is not the correct approach. If there are sufficient points of similarity, the apparent dissimilarities are unimportant.'

These two judgments are, in my view, misguided for several reasons. Firstly, they render the court seized with a matter vulnerable to the vagaries of incompetence and fraud on the part of the fingerprint expert. Secondly, in many instances fingerprint evidence may be dispositive and so the expert in the matter will be usurping the function of the court on what may well be the ultimate issue. The logic flies in the face of the very sound judgment of Lord Cooper in *Davie* v. *Edinburgh Magistrates*. Cooper had this to say: 'The parties have invoked the decision of a judicial tribunal and not an oracular pronouncement by an expert ... the duty of experts is to furnish the judge or jury with the necessary scientific criteria to the facts proved in evidence.' This approach was followed by Eloff, Kirk-Cohen and Van der Merwe in the South African case *Maritime and General Insurance Company Ltd* v. *Skye Unit Engineering (Pty) Ltd*.

Garrett and Neufeld examine the role of the judiciary in a number of significant cases. Courts policed the introduction of forensic testimony in these trials in a highly deferential manner, typically trusting the jury to assess the expert testimony. The courts were slow to keep even erroneous forensic testimony out, even on the comparatively few occasions when challenges were launched. In one instance, Joyce Gilchrist, who must enter the rogues' gallery as one of the most dishonest witnesses ever to disgrace the name of forensic science, fabricated evidence, yet the court was unwilling to reject her evidence. It took three attempts to get the evidence thrown out. The accused was finally exonerated with DNA tests in 2007 – after having served twenty-two years in prison. Courts denied relief to people who were subsequently exonerated despite their assertions, later shown to be correct, of flagrantly invalid forensic testimony. Judges did not remedy most errors brought to their attention.

A recent paper by Joseph Sanders, '"Utterly Ineffective": Do Courts Have a Role in Improving the Quality of Forensic Expert Testimony?', highlights the issues. The title of the paper reflects the levels of frustration experienced by this author. He points out the deeply rooted problems of a judicial mindset that results in decisions tending to preserve the official position and the status quo, and notes that, despite the vigilance displayed by courts in preventing junk science from entering the courtroom, judges are reluctant to throw out weak forensic evidence in criminal matters. I find this particularly interesting, as this mirrors my own experience in the South African courts. I have found consistently that although the burden of proof on the accused is legally a mere 'reasonable doubt', the true position for any accused wishing to challenge state forensic evidence is that he or she will have to prove the issue way beyond reasonable doubt, to the extent that the burden is for all practical purposes completely reversed.

One of the most disturbing suggestions to emerge from Sanders's paper is that 'judges are prepared to admit forensic science that fails to measure up to standards demanded by real science because it falls in line with preconceptions about the guilt of the accused'. This subject is also dealt with in *Strengthening Forensic Science in the United States*, where the authors say about the less scientifically rigorous aspects of forensic science, 'Some courts appear to be loath to insist on such research as a condition of admitting forensic science evidence in criminal cases, perhaps because to do so would likely demand more by way of validation than the disciplines can presently offer.' How sad.

I have seen various examples of judges' bias towards the state over the years. In a criminal case in KwaZulu-Natal, the 'expert' in the matter, Anthony Young, falsely claimed to have a diploma in analytical chemistry. The matter was referred to the Attorney-General and Young was charged with perjury. Later he was acquitted. By contrast, in the Van der Vyver matter, the perjury by two state experts went unpunished. Neither expert faced perjury charges despite their obvious lies and deceptions. Could it be that the disparity arises out of Young testifying for the defence and the other two for the prosecution? Similarly, the two quotes by Lord Denning towards the beginning of this chapter stand in marked contrast to each other. In the second, Denning gets it so right in preventing unjust harm to befall a citizen through administrative wrongdoing, yet in the first he seems to be quite complacent about the fact that six innocent men were beaten by police to extract confessions and that the forensic science used to convict them was not even good enough to be wrong.

It is surely time for the courts to recognise their essentially neutral role in the process of the administration of justice. It is surely also time that the courts re-evaluate the position in which an accused person has to prove his innocence rather than the state having to prove an accused's guilt. This not-so-subtle change in the goal posts is of concern not only in South Africa, where I have seen it in operation, but also in the US. As Steve Sheppard writes, 'The courts have moved the jurors' goal from a vote for the state if the state can convince them of a fact to a vote for the state unless the defence can convince them of a certain type of doubt.'

Chapter 18

'BUT DOCTOR, WHY WOULD THIS POLICEMAN LIE?'

'If the system turns away from the abuses inflicted on the guilty, then who can be next but the innocents?'

– Michael Connelly

'Every society gets the kind of criminal it deserves. What is equally true is that every community gets the kind of law enforcement it insists on.'

– Robert F. Kennedy

The chapter title is a question put to a witness who was challenging a police version of events. A flippant answer is, 'Just to keep in practice.'

Police and prosecutors and their ancillary support groups tend to adopt an 'us and them' attitude. The police in particular are a tight-knit group whose loyalty to the force and the badge easily transcends honesty, should the need arise. One need only remember the vastness and horror of the atrocities committed by the South African hit squads and the curiously named Civil Cooperation Bureau to know that the code of silence was maintained for decades.

Police promotions are driven by the number of arrests made, not by the justice and peacekeeping endeavours in which officers are involved. I remember a particular murder trial that took place in the Johannesburg High Court in the early 1990s. I was involved in the defence's case. A young man in his late twenties, a butcher by trade, had been having marital difficulties with his wife, who was a reserve policewoman and carried a pistol. One day, on arriving home after work, she told him that she wanted him out of the house. He persuaded her to let him shower and pack first. On emerging from the shower, an altercation developed and she produced her service pistol. A tussle for the gun resulted in a shot being fired. The bullet entered the wife's left nostril and killed her. The husband was panic-struck: here he was with a dead wife and a smoking pistol in his hand. When the police arrived, he made up a story, telling them about his wife's depression and how he had found her just after she had shot herself.

Nothing happened for some time – until his conscience got the better of him and he phoned the investigating officer and came clean. The investigating officer told him that his story would never be believed and then, in the friendliest way imaginable, he said to the gullible young man, 'Look, I have had considerable experience in this type of death. Just come with me to a magistrate and tell him you had an argument with your wife and you just lost control and shot her.' The investigating officer told him that it would go no further and that the courts would not convict him. 'You will be in and out in five minutes,' he said.

This is exactly what happened: the young man made what amounted to a free and full confession to a magistrate and then found himself charged with murder, the confession being the mainstay of the prosecution case. In those days the death penalty was still in operation.

After seeking legal advice, the accused's advocate advised him to challenge the confession, and the investigating officer went into the witness box and smoothly denied coercing the man in the manner described. This would have gone very badly for the accused had his brother-in-law not been sitting in court. At the tea adjournment, the brother-in-law approached us and confirmed that he had been present when the coercion had taken place. He validated exactly what the accused had told us.

Now why would the cop have lied in court? He did it because it was the lazy, easy way to secure an arrest and a conviction. Why did the prosecutor not inform the defence that the brother-in-law had disclosed the same information to him? For the same dishonest reasons, I am sure. In our law, as in many other legal systems, the prosecutor is legally obliged to disclose exculpatory information to the defence. They rarely do. In fact, my experience is that the prosecution usually conceals such evidence.

In the Van der Vyver trial, the prosecution kept from the defence the information that the crucial fingerprint had not been marked up at the scene of the crime, when it was lifted, but had been marked up some days later. In the case of a hotly contested piece of evidence such as this, this information is crucial, as evinced in earlier discussions of the matter. In the prosecution docket was an affidavit from the AFIS officer saying that the print sent to him was not marked as it should have been. What is even more concerning is that, as mentioned in Chapter 16, one of the prosecutors in this case approached the police forensic DNA specialist and put pressure on her to alter her report to say that she had found blood on the alleged murder weapon, when in fact she had not. It would seem that some prosecutors will do anything to secure convictions.

One of the major problems in dealing with prosecutorial conduct is that there is no accountability. It is rare that a prosecutor is charged for dishonesty in his or her conduct of a case. The reasons for this are quite clear. Prosecutors thrive in their own environment by winning cases, not by seeing to justice. Raymond Bonner investigated more than thirty cases in New York alone where the convictions had been overturned because of prosecutorial misconduct. In only one instance was a prosecutor punished in any meaningful way. Much more concerning was the fact that prosecutorial dishonesty often resulted in promotions and salary increases, even after their cases were overturned on exposure of their misconduct.

In an analysis of the Innocence Project cases (see Chapter 16), prosecutorial misconduct was found in 42 per cent of the cases. The long and varied occurrence of such prosecutorial misconduct suggests that the problem is systemic rather than isolated. The fact that most of these cases have come to light as a result of the retrospective use of DNA technology raises a serious concern that prosecutorial misconduct occurs across the board, but is merely less visible in the absence of DNA evidence, which has exposed so many miscarriages of justice.

'Shopping' for the 'right' expert is one way in which the system can become derailed. Fred Zain, Louise Robbins, Joyce Gilchrist, Michael West and Pamela Fish are all examples of experts who were sought out by prosecutors because they could be relied on to give testimony favourable to the prosecution; this, despite growing concerns about their competence and honesty. It is not good enough for a prosecutor to ignore these concerns. There must be an attempt to evaluate the evidence and to interrogate the witness instead of accepting blindly what they want to hear.

In the trial of Fred van der Vyver, the prosecution knew full well that the shoe-print expert, William Bodziak, completely disagreed with the police expert, Bruce Bartholomew. Not a single person in the field locally could be found to confirm Bartholomew's identification of Van der Vyver's shoe. Yet no attempts were made by the prosecutor to interrogate the evidence; the prosecution simply accepted Bartholomew's dishonest testimony and defended it to the hilt.

It is not only in this regard that prosecutors are dishonest. Prosecutors the world over are tardy in providing pre-trial discovery items. I am at present engaged in a matter where the prosecution possesses certain photographs. In the course of the hearing, the defence was given some of these photographs. When we inspected the police docket, it was clear that there were additional photographs in existence. On approaching the prosecutor, Mr Ntela of the Cape Town National Prosecuting Authority, he told me that he was not disclosing these additional photographs to the defence because he was 'not using them in the trial'. I should have thought it was obvious that this was added reason for the defence to see them, as they may have contained information helpful to the defendant. The prosecutor, while acknowledging that we were entitled to the photographs, said – and I quote – 'I am not giving them to you because I don't feel like giving them to you.'

We have seen this kind of behaviour often, quite apart from this particular prosecutor. In the Alan Clift matter discussed earlier in this book, it must have been painfully clear that Clift's original report contained vital exculpatory information that he chose to conceal from the defence. Where a situation is created that rewards prosecutors for successful prosecution only and provides them with significant power, yet demands from them no accountability, the scene is set for the encouragement of prosecutorial misconduct.

Part of the problem with prosecutorial misconduct is that, inevitably, it will be investigated by other prosecutors, and this is akin to having the fox investigate crimes in the henhouse. It is never going to curb the abuses. That this sort of misconduct results in wrongful convictions can be seen in an article by Peter A. Joy, 'The Relationship between Prosecutorial Misconduct and Wrongful Convictions: Shaping Remedies for a Broken System', in which he reiterates the point that prosecutorial misconduct is the result of three institutional conditions: vague ethical rules that provide ambiguous guidance to the prosecutors, vast discretionary authority with little or no transparency, and inadequate remedies for prosecutorial misconduct. In the US, the Model Rules of Professional Conduct govern prosecutorial conduct. However, their ineffectiveness can be gauged from Model Rule 3.3(a)(3), which allows a prosecutor to introduce evidence that is suspected to be false, or where there is a substantial risk of the evidence being false. Thus, a prosecutor may be reckless in introducing the evidence suspected of being false and escape censure. This begs the question as to whether wilful blindness of the prosecutor will invite censure.

While many aspects of the system function adequately, it is on the periphery of the whole edifice that we must focus our attention. In so many ways it is flawed and open to abuse. Forensic science itself needs to be re-evaluated. The '*CSI* effect' has crept into popular culture. It has also infected the law courts and imbued the lay mind (and I include the legal profession) with unrealistic expectations. The weaknesses of the traditional forensic methods have been starkly illuminated by the advent of DNA technology. Post-conviction investigations have uncovered alarming deficiencies in the way we do things. The standard police line-ups, where the accused and a number of other lookalikes are viewed simultaneously by the victim, are highly likely to produce misidentification. In the Innocence Project, by far the most prevalent factor in the wrongful convictions was incorrect eyewitness identification (75 per cent of cases). Cognitive psychologist Gary Wells has shown that eyewitness identification procedures using simultaneous line-ups produce a significant number of incorrect identifications.

The level of confidence in fingerprint identification has taken a backward step in the light of the Mayfield and McKie cases, where these were not minor blunders by low-level operatives but major blunders by the best of the best, all of whom claimed a zero error rate. The level of subjectivity in fingerprint identification is also concerning.

Ballistics or, more properly, tool-mark identification applied to bullets and shell casings, has likewise been shown to suffer from a weak scientific footing, as discussed in Chapter 15. Ultimately, the 'identification' is more subjective than scientific. Shoe prints, tyre marks and bite marks all fail to make the grade, and yet have passed muster in courts for many years.

Hair and fibre analyses have fallen off the forensic radar screen. The best that can be achieved with hair and fibre is to demonstrate class similarities; nothing more. The bottom line is that a great deal of forensic science lacks a proper basis in science.

Most real sciences are aware of cognitive biases that can invalidate the entire scientific endeavour. Standard scientific practice seeks to reduce or eliminate as much as possible the elements of cognitive bias. This is much less apparent in the forensic sciences.

Confessions are the life blood of many investigations. We now know that the standard Reid technique outlined in *Criminal Interrogation and Confessions* is instrumental in obtaining false confessions. Indeed, the interrogation process itself seems to generate false confessions. Between 14 and 25 per cent of confessions obtained using this technique may be false.

Of all of the above-mentioned flaws of forensic science in the system, no other single factor has contributed to the miscarriage of justice more than erroneous eyewitness identifications. This has been in the background of knowledge since Hugo Münsterberg's book *On the Witness Stand*, published in 1908, and, more recently, the writings of authors such as Elizabeth Loftus, who wrote *Eyewitness Testimony*. The psychology of identification and eyewitness testimony has not filtered down to forensic laboratories, police departments or the courts. The propensity of standard methodology to produce false eyewitness identifications is well known in academic circles. Sadly, there has been much resistance from officialdom. The police are resistant to ideas that do not emanate from within their ranks. The statement made by an American police officer concerning his belief in the law and the Constitution is typical and systematic: 'If we are going to catch these bad guys,' the officer said, 'fuck the constitution, fuck the bill of rights, fuck them, fuck you, fuck everybody. The only ones I care about are my partners.' (Mark Baker)

Well, you cannot put it more plainly than that! When you have that sort of attitude in the police, together with reluctance to learn from any outside source, there is a problem. A growing litany of carelessness, bias, incompetence, excessive cosiness with prosecutors and other embarrassing revelations have raised doubts about the trustworthiness and accuracy of some reported findings in a disturbing number of laboratories. Mnookin makes the point that 'the traditional forensic sciences are at this point inadequately supported by empirical data that would justify the strong claims analysts frequently make. We believe that numerous assertions made both in routine practice and in court are neither backed up by sufficient empirical data or research nor can these claims be justified or validated simply by reference to longstanding experience.'

Transparency and an ongoing critical perspective are absent from the police and state forensic service, not only in South Africa but also in the US and England, as I have shown in previous chapters. Paul Kirk, who in so many ways was a pioneer of the use of forensic science in law enforcement, pre-empted the issues that would come home to haunt the subject fifty years later. In 1963, he wrote:

> Forensic science progress has been technical rather than fundamental, practical rather than theoretical, transient rather than permanent. Many persons can identify the particular weapon that fired a bullet, but few if any can state a single fundamental principle of identification of firearms. Document examiners constantly identify handwriting, but a class of beginners studying under these same persons, would find it difficult indeed to distinguish the basic principles used. In short there exists in the field of criminalistics a serious deficiency in basic theory and principles as contrasted with the large assortment of effective technical procedures.

Despite the fact that this was written half a century ago, almost nothing has been done to remedy the position. What Kirk is saying is that this is all applied science, and not much attention has been paid to basic science. There are many reasons for the lack of experimental research and the lack of a research culture, and some of these are addressed in the next chapters. One of the reasons is a stubborn refusal to reconsider beliefs in the light of credible challenges. This is the very antithesis of the spirit and culture of the scientific method.

Chapter 19

BAD APPLES, BAD ORCHARDS OR BAD SOIL – OR ALL THREE?

'Talk unbelief, and you will have unbelief; but talk faith, and you will have faith. According to the seed sown will be the harvest.'

– Ellen G. White

In *Forensic Fraud: Evaluating Law Enforcement and Forensic Science Cultures in the Context of Examiner Misconduct*, Brent E. Turvey writes:

> The … majority of forensic examiners work for law enforcement or government agencies, and almost exclusively for the police and prosecution. Law enforcement culture is often defined by traits that afford the motivations and rationalizations for a deviant internal sub culture, actively cultivating fraud within its ranks. It also furnishes otherwise lawful members with the skills, incentives motivations and rationalizations for ignoring, protecting and even publicly defending their unlawful co-workers. Research suggests that employment circumstances and cultural features in

> conformity with Differential Association theory, Social Learning theory, and Role Strain [the strain experienced by an individual when incompatible behaviour, expectations or obligations are associated with a single social role] increase the likelihood that those aligned with law enforcement will commit, tolerate, conceal or defend acts of overt fraud.

In the same context Turvey goes on to suggest that there is institutional culpability for the high levels of forensic fraud within their ranks. In other words, he is suggesting that bad apples are produced in bad orchards. We have seen in previous chapters how the system is motivated and encouraged for the wrong reasons. This important theme is explored further in this chapter.

I would like to start with my own experience as a private forensic scientist in the South African context. In 1987, I became involved in a case that would achieve later notoriety as the 'Gugulethu Seven' case. The official version of events provided on national television initially went as follows: It was known that a vehicle carrying police officers to work was to be attacked by members of the ANC's armed wing, Umkhonto we Sizwe. According to the police spokesman, countermeasures were put in place and, when the attack came, the police were in position to thwart it. After a substantial gunfight, the score was no injuries of any consequence to the police and seven dead on the ANC side.

Eyewitness evidence, however, did not support this version. In particular, two locals who saw the whole saga unfold related a story much more in keeping with a police assassination of seven unarmed men. A local newspaper related the latter story, and one of the journalists, Tony Weaver, passed the story on to the British Broadcasting Corporation. Weaver was subsequently prosecuted for allegedly publishing false information about the police. This crime carried a substantial prison sentence.

At the trial, there were some interesting observations to be made.

Firstly, the official statements that were drafted by the police, under oath, were all clearly from the pen of the same author. The grammar, style and syntax all screamed one individual. Secondly, the statements were precise about who shot whom and with what. Unfortunately, the bullet wounds on the deceased simply did not match the events chronicled in the statements. One was left wondering whether the events depicted by the post-mortem reports were the same events deposed to in the affidavits. Thirdly, and I believe significantly, not a single state pathologist was called to testify. I have no doubt whatsoever that pathologists were consulted by the state, therefore I think there can only be one explanation for none of them testifying. They must all have known about the discrepancies in the affidavits. All of the pathologists involved later attended the second inquest to put up all sorts of improbable theories as to how the various indications of close gunshot wounds might have come about. It came as no surprise when the magistrate found that the police had no case to answer. There was, however, a twist in this tale. When, some years later, the truth emerged about hit squads and third-force activities, one of the participants in this squalid affair gave evidence before the Truth and Reconciliation Commission. The true events of that awful day back in 1987 emerged and confirmed my original findings.

The matter could not have been concealed for so long without the actions or tacit involvement of every facet of the system: the hit-squad members themselves, their friends and colleagues in the police force who were involved in the subsequent investigation, the prosecuting authorities who helped in the judicial cover-up, the pathologists who could see nothing wrong with the vastly divergent facts and their accompanying affidavits, and the magistrate who heard the matter in the most biased of ways. The individuals concerned were not just bad apples acting in isolation; they were nurtured in a bad orchard, which itself was grown in bad soil.

Thompson, in 'Beyond Bad Apples', dismisses the notion that these failures are simply the result of a few individuals suffering from moral turpitude and that the only activity needed is to weed out the bad apples. Indeed, no effort is made to find out why there are so many bad apples in the first place and why they seem to occur in clusters.

There are many examples, ranging from the poor quality of the bullet-lead comparisons in the top forensic science laboratory at the FBI and the FBI's failure to admit to the problems in the Mayfield case, to the Bronwich report on the Houston Police Department Crime Laboratory, which, as mentioned in Chapter 16, was closed in December 2002 owing to a scandal involving shoddy work and fraudulent reporting of findings. One would have hoped that when the laboratory reopened in 2006 things would be different. Alas, no. In 2008, the entire unit was shut down again and the new supervisor was forced to resign amid allegations that she had helped analysts cheat on mandatory proficiency tests.

Thompson argues that crime-laboratory failures can be most productively examined through the lens of organisational theory. This follows the work of Charles Perrow and Diane Vaughan, whose research focuses on institutional, social and cultural factors that shape human behaviour, and on the organisational environment in which they occur. Briefly stated, the criminal justice system comprises sub-systems such as police departments, courts, prosecutors, laboratories and forensic scientists. Within each sub-system are units such as a ballistics unit or a serology laboratory. If something goes wrong within a particular unit, it is described as an 'incident'. Incidents can range from unintentional mistakes through to deliberate fraud. In some instances, the incident can be so devastating that it leads to unit failure. A good example of this is a false report wrongly linking an individual to a crime. System failure results when there are multiple unit failures. Unit failure need not necessarily lead to system failure; in a good system there are checks and balances that provide what organisational theorists term 'redundancy' (in other words, belt and braces). To function well, the various components need to be loosely coupled, or independent of one another. It does not help to have the braces linked to the belt, because failure of the belt will result in failure of both, and down will come the trousers.

Unfortunately, the criminal justice system is not a loosely coupled system. As discussed in previous chapters, there is a tendency for prosecutors to overlook system failures in the forensic laboratories and there is a tendency for judicial officers (not all of them) to place greater store in state forensic evidence than in any expert testimony that discredits it.

This is exactly what we encountered in the trial of Fred van der Vyver. The deviant behaviour of the state forensic witnesses should have alerted the prosecution to trouble ahead, yet they failed to ask simple questions of their witnesses. If they had, the trial should have been stopped in its tracks. In fact, the prosecutors participated in the wrongfulness of the police behaviour and sought, by legal sleight of hand, to paper over the widening cracks. The judge failed to see the deficiency of the evidence, and so the whole misguided prosecution blundered on until an acquittal was achieved by an overwhelming mass of evidence indicating fraud and dishonesty in the state case. If Van der Vyver had not managed to assemble the formidable team that he did, he would undoubtedly have been sent to prison for a crime that could not be linked to him in any way, other than by the fact that it was his girlfriend who had been murdered. Unfortunately, this case is not unique to the South African legal scene. The wrongful conviction of sixteen-year-old Josiah Sutton in Texas shows just how wrong the system can be. In this case, there were multiple system failures, leading to catastrophic results: Sutton served almost five years of a twenty-five-year prison sentence for a rape he did not commit before being exonerated and released in 2004.

System failures reside both with the state and with poor-quality defence lawyers, who fail to prepare for and understand forensic evidence. As has been noted, the role of the prosecutors in the procurement of wrongful convictions cannot be discounted either:

prosecutors favour experts such as Louise Robbins, who can 'see' things that not even other government experts can see. This is no different from Bruce Bartholomew, who saw things in the Van der Vyver 'shoe print' that none of his colleagues could perceive. The list of corrupt forensic experts is long (see, for example, the list in Chapter 16) and includes those knowingly selected by prosecutors despite being openly corrupt. Fred Zain was one such expert: the prosecution continued to use him despite incontrovertible evidence that he was not only incompetent, but a fraudster to boot.

Zain was hired in 1977 as a chemist in the West Virginia State Crime Laboratory. It was subsequently established that he had no scientific qualifications whatsoever and, despite having failed an FBI course in forensics, he was hired; clearly no one had checked his credentials. Although Zain failed to pass any further courses, he was retained and held in high esteem by the prosecution because of his ability to 'solve' cases that confounded other forensic scientists.

In later years, and after a string of trials where his evidence was used, it was discovered that Zain was an out-and-out fraudster. The results of the investigation were so appalling that it was held by at least one High Court judge that any evidence offered by Zain should be presumed prima facie to be 'invalid, unreliable and inadmissible'. The post-conviction application of Glen Woodall states: 'The matters brought before this Court ... are shocking and represent egregious violations of the right of a defendant to a fair trial.'

No attorney can adequately run either a civil or a criminal case in which forensic evidence is tendered without a firm grounding in such evidence. In my view, all law degrees should have a proper instruction course in forensic science. Most courses have training in forensic medicine, but this is wholly inadequate to equip a lawyer for the purpose of cross-examining an expert in the fields of, say, handwriting, fingerprints, DNA, fires and the host of scientific matters that come under judicial scrutiny.

As for forensic science itself, there is now debate as to just how scientific the subject is. As discussed in previous chapters, there is a growing sense of discomfort by some members of the scientific and legal communities that some of the grounds on which the actual science rests are flimsy and, in some instances, non-existent. In an insightful paper, Gary Edmond of the University of New South Wales, states:

> To the extent that forensic science and medicine have been historically insulated from more mainstream scientific and biomedical research, imposing expectations that require evidence of testing or other indicators of reliability would seem to be important responses that will assist the courts as well as the investigative institutions and laboratories to improve the standard of expert evidence relied upon in criminal prosecutions, convictions and appeals. Moreover, the state is in a position to take remedial steps to ensure that expert forensic evidence is subject to a variety of testing and validation procedures.

Edmond goes on to say that the majority of literature on law, science and medicine 'appears to be oblivious to decades of research by historians, philosophers and sociologists specializing in the study of science, medicine and technology. This is unfortunate because this latter body of work challenges many of the conventionally held views about expertise routinely and somewhat glibly employed in legal discourse practice and proposal for reform.' Under the title 'Forensic Medicine and the Forensic Sciences as Law–Science Hybrids', he writes:

> Most of the fields we are discussing did not grow out of basic science ... There is no systematic rigorous, empirical research on which the forensic identification sciences knowledge is built. If called upon to prove their claims, they have little or no data to marshal in their support. Instead [they] point to a guild of mutually self-reassuring examiners who have come to believe in the truth of their claims, often sounding more like a faith-based religion than a data-based science.

Much of the forensic testimony that I have encountered relies heavily on the 'experience' of the witness. Experience is of little assistance in debating scientific results or scientific conjectures and hypotheses.

Michael Sax states, 'Forensic science and medicine have evolved in a symbiotic relationship with the criminal justice system. From the judicial perspective that relationship has been characterised by trust rather than scrutiny or accountability.' This latter paragraph may go some way to explaining the actual situation in court. It has always been my experience that it is far more difficult to get a criminal court to accept my evidence than a civil court, despite the fact that the burden of proof for an accused is so much less burdensome than the prevailing situation in a civil matter. It would seem that I am not alone in this. Despite vigorous beginnings in *Daubert* – the judgment that urged judges to be gatekeepers and to keep junk science out of their courts – judges have failed miserably in this role in criminal cases, despite having carried out this function assiduously in civil matters. The close relationship between forensic sciences and prosecutors seems to have produced a climate of pro-prosecution sympathies.

Edmond writes:

> Forensic science and medicine tend to be applied sciences. In practice, such an application is not particularly meaningful. It certainly does not provide an excuse to circumvent the need for rigour, especially the need for empirical validation to demonstrate reliability and individual competence ... Historically, forensic science and medicine have relied upon 'art' and 'experience' in addition to experimental techniques, which can hardly be said to be reliable and should be regarded with great suspicion and circumspection by judges. Unfortunately, this is not often the case ... It is surely time to wean the courts away from blind faith in the state forensic services.

Confidence in the state forensic science laboratories is not a substitute for evidence of reliability. It is important to remember that judges should be concerned with evidence of the reliability of particular techniques and theories, not evidence of the eminence of scientists, their performance or their credibility. Judges should no longer take the reliability of evidence generated by the institutionalised forensic sciences on trust.

As mentioned, as long as we have a police system incentivised on the basis of the number of arrests made and as long as the prosecutorial system's primary goal is the attainment of convictions, the seeds for miscarriage of justice and corruption will have fertile soil in which to grow. The social gulf that exists between the police and 'the other side' leaves no room for any interactive learning. The police frequently see themselves as the 'good guys' who make the world a safer place in which to live. They socialise primarily with other policemen and, interestingly enough, also with the criminal underbelly of society. Because of their one-sidedness, many people within the police force are completely unaware of their cognitive biases. They do little or nothing to guard against them by using strict procedures and protocols specifically designed to minimise such biases. Despite mountains of scientific evidence relating to police line-up procedures, those governing identification procedures have steadfastly refused to take cognisance of the ways to improve the work they do. Despite Itiel Dror's work showing the subjectivity of the fingerprint identification process, the concepts have not gained much traction with the authorities. Not until 2011 did the standards-setting body for identification create protocols for blind verification. What is more concerning is that the protocols that they did create were not applicable to every case and, in any event, were not binding.

Why is it that the legal system, including judges, prosecutors, forensic staff and police, are all so resistant to changes that would make the system work more fairly and more accurately? It cannot be that prosecutors and judges are happy to countenance miscarriages of justice.

I have already commented on the concept that a system that rewards the wrong things (arrests and successful prosecutions) can never avoid dishonest practices and careerists who pay lip-service to the concept of justice and pursue a prosecution at all costs. This seems to be endemic worldwide. As I write this book, a complaint against a senior member of the National Prosecuting Authority is being investigated. The allegations are briefly as follows: During the bail application proceedings, it became evident that a senior police official deposed on affidavit to the 'fact' that they had observed the crime scene (a boat), and that the video clearly showed the accused at the scene. It later transpired that the police officials had lied under oath about the name of the person who had filmed the video and about what could be seen in the video. The police became aware of the falsehood early on in the investigation and suppressed the truth, in conflict with their positive duty to investigate and disclose. Both senior advocates for the state also became aware of these facts at an early stage and likewise remained silent, failing to disclose the misrepresentation – which was material – to the defence or to the court. It also appears that an important alibi witness was intimidated by tax officials two days after his name had been mentioned in court and, moreover, that these officials were in possession of a document that had been handed in to court and appeared to be central to their inquiries and threats. It is difficult, if not impossible, to escape the inference that the prosecution collaborated in or, at the very least, were aware of, these dishonest activities, allowing misrepresentations to be perpetuated to further their efforts to have bail denied. Admittedly, the alleged crime in this case is a serious one (cocaine in significant quantities was found and seized), but in my view this does not condone dishonesty on the part of the police and the prosecution to pervert the course of justice.

In this matter, the rationalisation for these actions is not difficult to fathom. I interpret the rationale as: 'We as police and prosecutor must save the country from the drug scourge by any means possible. What harm can come from a few dishonest statements in the greater scheme of things, which is to get these guilty people off the streets?'

Here you have the beginnings of a miscarriage of justice. The prosecutors and police have already decided on the guilt of the accused and will make a conviction stand even if it is predicated on lies. In the middle of the last century, Leon Festinger introduced the notion of cognitive dissonance, which occurs when a person holds two conflicting beliefs. The behaviour of the state in this case may be explained in terms of cognitive dissonance. We have a prosecutor who sees it as her duty to act as the agent of good and, in the course of this activity, she has to lie and cheat. The only way out is to rationalise the latter and submerge it in the notion that the end justifies the means, and that the end result will be a good one despite the crooked path used to get there. This sort of behaviour played a role in many of the travesties that came to light in the Innocence Project.

In cases where evidence is challenged or shown to be wrong, the state will hardly ever accept the error, but will resort to more and more complex rationalisation to try to make the evidence stick to the point of irrationality. Take for example the so-called shoeprint in the trial of Fred van der Vyver. Bruce Bartholomew made his 'identification' on the basis of three, subsequently four, grains of sand in one of the grooves of the sole of the shoe. Against standard procedure, he did not make an inked print of the shoe. The reason given at the trial was that he 'was concerned lest the inking of the shoe and application of the sole to a sheet of paper should dislodge the grains of sand'. No blood was found on the shoe, which means that it would have undergone a thorough scrubbing and washing. Yet this did not dislodge the sand particles. Bartholomew's solution was to ignore this uncomfortable fact.

To make the mark in the bathroom, the shoe must have come into contact with the blood of the deceased on the floor. Yet no bloody marks were observed between the lounge, where the murder took place and where the blood was found, and the bathroom. The prosecutors in this matter were aware of these impossibilities. They were also aware of the disparity between what Bartholomew had told them and what William Bodziak was telling them, yet they rationalised it all away, hoping that Bodziak would not be called by the defence. The cognitive dissonance generated was dealt with, again, by deceit and rationalisation, and the simple failure to apply their minds to interrogating their own evidence properly. Had they done so, they would have seen the weaknesses and would not have sailed on regardless.

In the opening address to the court, the prosecutor told the judge that he did not intend to rely on the fingerprint (which, you will remember, was pivotal to the case). What more doubt about the evidence do you need, given that the prosecutor had enough doubt to verbalise it in his opening address? Yet the case proceeded at staggering cost to the accused and the evidence was always pathetically weak. The relentlessness with which the prosecutors conducted themselves was encouraged by the fact that there is no accountability in the system. Had there been some sort of costs award against the state or the prosecutors, a different course of behaviour would surely have been demonstrated. The same applies to the police and to the forensic experts in the matter.

Alan Dershowitz, the high-profile lawyer in the Claus von Bülow murder trial and a professor of law at Harvard Law School, ignited a veritable firestorm when he said on national television that 'police departments tell their detectives it's okay to lie, they learn it in the academy'. It is ironic that a commission specifically set up to investigate corruption in the New York Police Department (NYPD), the Mollen Commission, found not only that the NYPD was beating up and robbing drug dealers to resell the confiscated drugs, but that lying under oath was widespread and condoned within the police department. It was even given a name: 'testilying'.

Former police officer Michael Dowd did not just take bribes from drug dealers to turn his head; he became a drug dealer himself and actually assisted and protected major drug operators. Former police officer Kevin Hembury not only stole drugs, gems and money in the course of a series of unlawful searches, but was part of a gang of cops that raided drug locations almost daily with the sole purpose of lining their pockets with cash. Former police officer Bernard Cawley was nicknamed 'the mechanic' because he openly and frequently 'tuned people up' – in other words, beat them up. To cover up corruption, the officers created false reports and committed further perjury to conceal the misdeeds within the NYPD.

Sadly, this is all too familiar. How do the various parties deal with the cognitive dissonance created by these actions? Most of it is rationalised away, but in some cases, such as the ones referred to above, I believe that even rationalisation cannot reduce or conceal from the corrupt individuals the wrongfulness of their actions. There develops an arrogant sub-culture that causes these individuals to believe that they are beyond the reach of the law, or that, somehow, the law does not apply to them.

Much of the rationalisation has to do with the conviction that they are doing 'God's work', preventing criminals getting off on technicalities. Rationalising away the cognitive dissonance is almost certainly the basis of the resistance to changes in the forensic science paradigm at all levels of the justice system. In South Africa, the fact that the political system under apartheid condoned corrupt and antisocial behaviour among the police complicated the issue, as everything could be justified as supporting the prevailing ideology. More recently, the system has again become skewed by the corruption filtering down from the highest office in the land. The president appointed corrupt party cadres to head the police and chose key elements within the justice system to obstruct the legal process from catching up with his own corrupt activities.

Many factors lead to a dysfunctional system. The attitudes of prosecutors who see themselves on a crusade for truth and justice are not helpful. The problems referred to in this chapter run deep, and I shall deal with potential remedies in Chapter 21. However, one thing needs to be stated very firmly. In all the cases of forensic science fraud, incompetence and malfeasance that surfaced in the Innocence Project, not one came to light as a result of aggressive adversarial legal intervention by the defence. The notion that the courtroom can serve as a 'crucible' where the truth will inevitably out is pure mythology.

Chapter 20
A TALE OF TWO CULTURES

'Rather fail with honour than succeed by fraud.'

– Sophocles

In South Africa, the police laboratories operate in a climate of secrecy. Thirty years ago, I was informed that the head of the state laboratory, the infamous Lothar Neethling, had compelled the entire staff to sign a document saying that they would have no dealings of any sort with me. They were not even allowed to talk to me. Once Neethling had left the laboratory in the more capable hands of General Heinrich Strauss, the rule was relaxed, but only after some anxious moments on the part of Strauss, who came from an academic background in organic chemistry at the University of Pretoria. Strauss's openness towards me, as opposed to the paranoid behaviour of his predecessor, is a reflection of their different backgrounds – academic versus police.

This dichotomy in the two cultures is dealt with extensively in the PhD thesis of Brent Turvey quoted in Chapter 19. The majority of forensic staff is qualified only to a level that I would classify as a technician. The contrast between a technician and a scientist is huge. Mostly (and there are exceptions) technicians are trained to a lower level of theoretical competence than postgraduate scientists. Generally technicians are employed to do repetitive, routine activities often requiring great precision, accuracy and skill. Making measurements is one thing; interpreting the meanings of these measurements is something entirely different. The problem is compounded further by the fact that forensic scientists in South Africa and in other jurisdictions are employed in a quasi-military framework where orders are given and must be obeyed even if they conflict with the scientific ethic of the individual.

The scientific culture is defined by the scientific method, which enjoins strict prohibitions as well as strictly defined and accepted practices. That is not to say that they are always obeyed. One only has to read the book *Betrayers of the Truth* by William Broad and Nicholas Wade to get a glimpse of the world of scientific cheating. Interestingly, and not surprisingly, the same features that engender fraud in forensics are present in the more rarefied climate of academe. Scientists are human and display all the frailties of the human condition. One would expect some situations in science to produce a culture ripe for fraud. Normally, if the work is important enough, attempts will be made to reproduce it, and that is where the whole pack of cards usually comes tumbling down.

Science acts in this self-policing way for a number of reasons. There is or should be a culture of intellectual sharing. Collaboration is a desirable activity in the world of science – the scientific journals are open to all to read. By and large, most laboratories will welcome fellow scientists and will usually be more than willing to show and tell. Integrity is crucial to the scientific endeavour; there is no point in doing research if fraudulent methods are involved. As a scientist, the primary goal is to try to model the truth as closely as possible. Science involves the ethical values of honesty, trustworthiness, openness, objectivity and fairness. It also embodies the notion that scientific findings are not set in stone, nor are they immutable truths.

If one examines the context in which law enforcement operates, one finds that it is seriously conflicted with the realities and ethics of the scientific endeavour. The secrecy with which state forensic science operates is in direct conflict with the openness of science. The aggressive quasi-military environment is inimical to the culture of science, which demands analytical logic and the consideration of alternatives. As made clear in Chapter 19, the tolerance of corruption and dishonesty is condoned in the world of law enforcement, where the operatives believe they are doing God's work against the forces of darkness, so what does it matter if a few lies are told to get the evil ones off the street? There is, of course, no place for this in science, and transgressions are eventually weeded out, the perpetrators banished, never to reappear on the scientific stage.

By and large, fraudsters in the scientific arena are exposed by colleagues, often with a degree of *Schadenfreude*. In police culture, loyalty to the group is paramount – the appearances of integrity must be kept up at all costs. One has only to look at the FBI press release in the Mayfield case to see this (see Chapter 6).

An indication of the gulf separating the two cultures is illustrated in this table from Turvey's thesis:

Law enforcement culture vs Scientific integrity

Law enforcement culture	Scientific integrity
A 'noble cause' belief system/ varied toleration for corruption	Zero tolerance for misconduct
Authoritative/coercive	Logical, empirical and fair
Masculinity and aggression	Humility re: limits of findings
Group loyalty/solidarity	Critical thinking/scepticism
Deception as a viable tool	Honesty in reporting
Isolation/'us vs them'	Openness to peer review and independent validation

Secrecy/'code of silence' towards error and misconduct	Transparency of errors/report all misconduct
Punishment/ostracism for those who break the 'code of silence'	Protection for whistle-blowers
Project and protect image of professional integrity	Project and protect image of professional integrity

It is detrimental to the entire enterprise of truth-seeking for the state forensic scientists to consort only with the prosecution. Stupid errors are made and these wreak havoc with people's lives. Yet, as it will have emerged in the discussions in this book, the official forensic scientists involved on the state side will not discuss the problem, preferring to busk out the evidence in court in front of magistrates and judges who, in many instances, have a bias in favour of the state institutions. (Many, if not most of the magistrates who hear the vast majority of criminal cases, have come from the ranks of the prosecutors.)

As we have seen, the problems are caused not just by one or two bad apples, but by a bad orchard, which encourages the growth of bad apples. Multiple examples of fraud and poor scientific work throughout the system are evidence of wilful institutional and supervisory negligence, if not ongoing cultural pressure to commit acts of fraud.

It is my view that any institution that has a spokesperson as its only conduit for communication with the public cannot have a commitment to honest communication. The chain of command in the police forensic system prevents honest officials from voicing disquiet anywhere but in-house. This will inevitably dilute, sanitise and, in most instances, conceal the truth from the wider public. If employees break the rules and speak out, the punishment is both swift and dire and includes suspension and termination of employment with loss of wages, medical benefits and pension. Restrictions of free speech are contrary to the spirit of scientific endeavour.

A good example of the harmful effects of the suppression of information has recently been revealed. Unfavourable task-force findings regarding forensic casework under review since the late 1990s have been intentionally suppressed and concealed by the US Department of Justice and the FBI, thus harming an untold number of criminal defendants. In South Africa, there is the same reluctance of the state to admit to any problems. In the police forensic laboratories, senior officers have either left, or remain as demoralised groups who would leave if they were not trapped by the fear of the realities of the commercial world, when all they have known is sheltered employment.

While conceding that there are major problems in the world of forensic science, the National Academy of Sciences' report is rather reticent about actual fraud among practitioners. As I have illustrated in this book, the second most common cause of miscarriage of justice is faulty and, in some instances, fraudulent forensic science. There is clearly a need to foster a research culture in forensic science, and forensic scientists should focus more on getting the science right than on winning cases.

Little has been done to evaluate the influence of cognitive bias on forensic results and, similarly, contextual and conformational bias is widespread and poorly researched.

Our courts should back away from evidence based on expertise and experience unsupported by scientific data. This will focus the judicial mind firmly on evaluating the evidence and not the witness. Ultimately empiricism is one of the cornerstones of any research. As illustrated, far too much emphasis is placed on experience and far too little on empirical verification of the results. Experiments that have deep methodological flaws in them should have no place in court. The scientific community is thoroughly indoctrinated with the notion that all ideas, paradigms and theories are provisional. Forensic science, however, does not present itself this way in court. Blind proficiency testing is the exception rather than the rule.

Casework can never take the place of structured research and, ultimately, never goes beyond the anecdotal. There is an almost desperate need to raise the academic educational level of a significant number of the forensic scientists in everyday practice. With this should come more sophisticated thinking about data and results, and especially a culture of questioning the legitimacy of inference.

In general, it would be helpful to have a culture of university-based forensic training, which could function symbiotically with the daily forensic routine practice. That way, one might conceivably produce a true hybrid species, and not the technicians that we have now, with their limited research experience.

The availability of a research-based academic forensic community would provide resources to the private legal fraternity. In my view, it is essential that the forensic playing fields be levelled. In recent communications, both personal and in an address to the fledgling Forensic Science Society in Cape Town, the course leader and coordinator disclosed that she was encountering significant resistance from her medical colleagues to her provision of forensic input into Legal Aid defence cases. This merely emphasises the points already made in this book. Such behaviour is not scientifically acceptable, it is not in the spirit of justice and, unfortunately, it displays the soft underbelly of forensic medicine: in my opinion, bias and lack of insight on the part of the senior medical members of the Department of Forensic Medicine at UCT.

The paranoid secrecy of many forensic laboratories does not fit into the framework of science. The accessibility of journals such as that of AFTE should not be reserved for the sacred initiates to the subject; all information should be available to all scientists and, for that matter, to anyone with the requisite knowledge or interest.

In addition to all of this, there is almost no regulation of forensic science. In any medical chemical pathology laboratory there is a quality-control procedure in place. This involves blind testing of known samples and the ongoing revision of all procedures in the laboratory. Not so in forensic science. In medicine and dentistry any would-be practitioner cannot practise without being suitably qualified to register with the appropriate council. While the medical forensic pathologists have to register, this is not true for forensic scientists. Anybody can hang up his or her plate and practise as a forensic scientist. This seems to be an international problem.

The field of forensic science is crying out for some academic and professional regulation. The realisation that quality control is vital to the proper running of the forensic science laboratory has been with us for some time. L.W. Bradford wrote about barriers to quality achievement in forensic laboratories as early as 1980. The results of his deliberations are frightening, as the table below shows.

Unacceptable responses

Category	Sample number	Labs submitting unacceptable responses (%)
Blood identification	3	71
Paint identification	10	51
Soil identification	11	36
Glass identification	9	31

The abysmal results in this case were ascribed to failure to apply adequate and appropriate methods. It is also noted that there are no codified standards of practice.

Concerningly, nothing seems to have changed a decade later. In a paper written in 1991, Randolph Jonakait provides an avalanche of evidence about just how sloppy the laboratories were. The poor quality of the work and errors in conclusions ranged over all aspects of the forensic repertoire and over all the major laboratories polled.

In one case, various laboratories were sent three 0.25-calibre bullets, two of which had been fired from one weapon and one from a second weapon. Of the responses, 9.1 per cent of the conclusions were just plain wrong. In the case of bloodstain analysis, only 28.8 per cent did the analysis correctly. In the case of paint comparisons, only 48.9 per cent of the laboratories got it right. In the case of handwriting comparisons, only 45 per cent of the forensic document examiners reached the correct finding. The authors of this study conclude that 'the kindest statement that we can make is that *no available evidence demonstrates the existence of handwriting identification expertise*' (emphasis added).

Yet, despite all of this hard data, junk science continues to pervade the courts. The failure to educate forensic scientists adequately, the failure to regulate the subject and the prosecutorial mindset of many of the practitioners are responsible for this sorry state of affairs. Although there is a growing number of academic forensic science courses, the products of these courses have yet to make an impact on the problem. In court, the absence of standard protocols in forensic laboratory work is concerning in the extreme. Yet forensic scientists defend this deficiency by asserting that standard practices can only be used with standard samples and crime laboratories do not analyse such samples. Benjamin W. Grunbaum, writing about 'potential and limitations of chemical analysis of physiological evidence', states:

> in many areas of scientific endeavour 'official' or 'standard' methods are used that are more conducive to rigid adherence than to improvement. Most improvements in methods evolve from a series of minor modifications to an original procedure made by persons of ingenuity and flexibility ... Forensic scientists are reluctant to use the standard or official approach preferring to use instead the methods they themselves have validated.

This approach, in my opinion, is flawed. Firstly, it results in everybody developing their own quirky methodology that cannot be evaluated easily by others. Secondly, it is highly unlikely that the variations introduced into the method will ever be properly evaluated by peer-group verifications. As Jonakait writes, 'this approach is simply bad science'.

I am presently evaluating several sets of results of analyses of blood alcohol in persons charged with DWI. The results should be in duplicate for each accused and should differ, as mentioned in previous chapters, by no more than 5 per cent. Unfortunately, this seems to be unachievable by the laboratories – there are differences between duplicates of up to 127 per cent. The measurements of sodium fluoride, which is used as a preservative, are also important, yet I have results of a 23 per cent concentration of sodium fluoride. This is rather astounding, as sodium fluoride–saturated solution contains about 4 per cent of the salt. This deplorable situation arises as a result of political interference in the staffing policies of the laboratories. There is regrettably ever-mounting pressure on the universities to dumb down the courses that they offer, and a significant number of staff in governmental laboratories is drawn from this output. The results are there for everyone to see. This political interference is superimposed on the already ailing system, resulting in a drop in standards.

Although the problems besetting forensic science are well known, probably more so outside the field than inside, not much has been done to remedy the situation. I am not so naive as to believe that the whole rotten edifice can be mended at a stroke. While I view the situation in the same light as the Augean stables, I have no great expectations that a latter-day Hercules will be able to divert a river and clean them. The process will take time and will require input from all players, not all of whom will be that enthusiastic.

The first step is to educate the courts in not blindly accepting state-based forensic science. Our judges and, indeed, the entire legal profession must be given the educational tools to evaluate the science rather than the witness. One of the crasser members of the Johannesburg bar always greets me with, 'Hello, Dr Klatzow, and in what are you an expert today?' The asininity of the greeting should be clear to anyone who understands Louis Pasteur's observation that 'there are no applied sciences, only the application of science'. Often the same scientific principles can be applied to a variety of problems. Once the bench and bar become more discerning about what constitutes good science, and especially about what inferences can be legitimately drawn from the data, we will have the first powerful influence to improving the science. This may entail specialised courses for judges and lawyers and may even require specialist judges to hear matters where a significant scientific content is being debated.

The second thing to be done is to remove all forensic services from the control of the police and the prosecution. Scientific and medical assessment conducted in forensic investigations should be independent of law-enforcement efforts to prosecute criminal suspects or even to determine whether a criminal act has been committed. Administratively, this means that forensic scientists should function independently of law-enforcement administrators. The best science is conducted in a scientific setting as opposed to a law-enforcement setting. Because forensic scientists are often driven by a need to answer a particular question related to the issues of a particular case, they sometimes face pressure to sacrifice appropriate methodology for the sake of expediency.

Recognition of the potential for bias and partisanship in state-run forensic laboratories has been recognised on the other side of the Atlantic. According to a British court in *R* v *Ward*, 'Forensic scientists may become partisan. The very fact that police seek their assistance may create a relationship between the police and the forensic scientists, and the adversarial character of the proceedings tends to promote this process. Forensic scientists employed by the government may come to see their function as helping the police.' Other commentators in the UK have said that 'the miscarriages of justice in England give unequivocal evidence that the pro-prosecution orientation of government scientists has not been countered in England' (see Freckleton, page 107).

If the laboratories are made truly independent, so that the prosecution as well as the defence have access to both the scientific expertise as well as the analytical facility, a much more level playing field will eventuate. To check any scientific work, there is a need to scrutinise the raw data. This is often problematic. Interestingly, it is not only within the forensic world that raw data is so inaccessible; this phenomenon also occurs in the wider scientific establishment. In some research done on this subject, a graduate student at Iowa State University wrote to thirty-seven authors of articles in psychology journals requesting the raw data on which the papers were based. Only nine gave unrestricted access. Twenty-one wrote back to say that the raw data had been lost or misplaced. Really! The nine sets of data obtained were analysed. Three of the seven that arrived in time for the analysis were found to contain gross errors.

Most forensic reports are very sparse on detail, which leaves the forensic official lots of wiggle-room in court. Comprehensive reports should be the rule and not the exception. The state forensic pathologist or scientist should be available for consultation with both prosecution and defence, which will result in a major saving in court time, avoiding the situation where experts argue about things where there may be substantial agreement. Yet in all conversations to date with the heads of the major forensic pathology departments, this notion has been firmly vetoed. Why?

These notions of forensic independence are not new. Meanwhile, there has never been a more acute need for forensic vigilance than the present.

EPILOGUE

I have tried, throughout this book, to illustrate some of the realities of the law, particularly where it interacts with science. We have seen that the role played by science in developing a case against the accused and in influencing many convictions is of more than passing importance. This notion is supported by research done by Joseph L. Peterson, John P. Ryan, Pauline J. Houlden and Steven Mihajlovic. In analysing the role and effect of forensic science on the outcome of criminal prosecutions, they found that about 25 per cent of jurors who were presented with scientific evidence would, in the absence of such scientific evidence, have delivered a not-guilty verdict rather than a guilty one. In addition, forensic science reports have a significant impact at the time of sentencing: convicted defendants are more likely to go to prison and for significantly longer periods in cases where scientific evidence is presented.

This book illustrates the fragility of the whole edifice of forensic science. At the turn of the twentieth century the personality cult of one man, Bernard Spilsbury, could influence the outcome of a trial, and we have seen just how wrong he, and others, could be. More recently, with the *CSI* culture upon us, unrealistic expectations have been generated not only with the layperson but also with jurists at every level. It is important to educate all those who might be end users of forensic science.

As we stand now, in 2014, forensic laboratories do not have to be accredited and most do not have external quality-assurance programmes in place. Indeed, forensic science has 'often treated the topic of quality control with hostility', causing concern that extends back to 1974 and an article by M.A. Thomson.

It is unfortunately true that the laboratory tasked with measuring a blood-glucose level is far more quality controlled and regulated than most forensic laboratories in the world, where the life of the individual quite literally might be at stake. We need to inculcate a better scientific understanding in all sections of the legal system and, as I have demonstrated, we must move away from evaluating the witness towards evaluating the evidence.

Every scintilla of scientific evidence that is put before a court requires careful scrutiny. It is crucial that the judiciary moves away from the pro-prosecution stance to establish, once again, the golden thread that should pass unbroken through every justice system, namely the presumption of innocence.

APPENDIX A

Prosecution address to the jury from Mr Justice T.R. Morling's report (Extracted and copied from the original)

'What I can say to you is this: that you've got Mrs Kuhl, who in her routine daily work as a forensic biologist using anti-serum tested to her satisfaction and to the satisfaction of Doctor Baxter, doing standard tests using standard techniques on old blood, managed to get 22 positive results for foetal haemoglobin. Which said to her that she was dealing with the blood of a child under 3 months old. You've got Dr Baxter, another government employee, checking the work of one of his biologists and he agreed with her 22 times, and you've got Mr Culliford, totally detached from it all, checking her notes, her evidence, her work methods, and agreeing with her 22 times. Now clearly this is very prejudicial and embarrassing evidence to the accused. Clearly it is damaging to them to have a car with traces of foetal blood in it; to have a camera bag with foetal blood on it, and I notice that it's not suggested how that might have got there. Not even Mr Lenehan can be blamed for the blood found on the zipper because the bag was only acquired 3 months before the event. I'm not going to go into Joy Kuhl's findings in detail because you remember them. You remember where she found foetal blood. You remember the scissors and the inside of the chamois container. You remember under the dash – I'll take you to that in a minute. You remember under the passenger's side seat, the 10 cent coin, the stains around the bolt hole, all of which have been photographed. You remember the stains down the vinyl of the passenger seat. You remember how she simulated similar bleeding to show how it happened. You've got the photograph there, how you have to sit on the seat for the blood to flow down under the hinge, and you remember the blood that she found under the hinge. Now you'd think, wouldn't you, that this would require a great deal of explaining even if it's adult blood. She says it is foetal blood, and I suggest to you that she ought to know, and Dr Baxter ought to know what it is he's dealing with, because you know really, if the suggestions made about their work in this court have any substance, people in New South Wales are in constant danger of being wrongly convicted whenever there's some blood involved, and it's really, I suggest, rather too ridiculous to contemplate that she would come to this, in the course of her daily work, as a professional forensic biologist, and muck it all up not knowing whether she was dealing with adult blood or the blood of a child under 3 months of age. What we ask you to do is respect her opinions.'

APPENDIX B

COPIED FROM LETTER
(*'s indicate illegible text)

BEHRINGSWERKE AKTIENGESELLSCHAFT
TELEFON MARBURG (0 64 21) 20 21
FERNSCHREIBER 04 *2 ***
TELEGRAMME BEHRINGSWERKE MARBURGLAHN
POSTSCHECKKONTO FRANKFURT /W. *** **.***
BANKKONTO COMMERZBANK AG. MARBURG 1
NR *** **** (* LZ *** *** **)

Behringsewerke Aktiengesellschaft Pos**ach 11 ** - D- **** MARBURG 1

**** ZEICHEN	IHRE KACKRICHT VOM	UNSERE ZEICHEN	DURCHWAHL 39	D-**** MARBURG 1
		Dr. Bd/Ln	(0 64 21) *** / 4255	21 Juli 1983

Mr.
Stuart Tipple
2 Baker Street

Gosford, NSW
2250

Manufacturing of Antiserum against Haemoglobin F/Batch-No. 2456

Antiserum against human haemoglobin F has been produced by Behring-werke since about 1971. It is produced by injecting a number of rabbits with purified haemoglobin F. The rabbits produce antibodies to haemoglobin F in their bloodstreams. After six to eight weeks selected rabbits are killed and their serum is collected and pooled after testing for activity and specificity. To remove unwanted antibodies in the serum if necessary, absorptions are carried out. For instance absorption with haemoglobin A is performed to remove antibodies to the alpha chain, which is common to haemoglobin A and haemoglobin F and also with human plasma. After completing quality control procedures, stabilization and sterilization, the pool of absorbed antiserum is given an identification batch-number, such as 2456.

The antiserum against haemoglobin F is then dispensed into bottles, each containing a 1 ml portion, which are frozen and maintained at a temperature lower than -20°C. From time-to-time, depending on demand, bottles of antiserum are supplied for use. Bottles of antiserum with the same identification batch-number, contain identical antiserum.

Our company records evidence that bottles of antiserum against haemoglobin F from batch number 2456 were supplied to the Australian distributor from April 1975 to September 1982.

Antiserum against haemoglobin F is not listed in our commercial catalogue since it is produced as a special laboratory product which does not have defined uses. Therefore, the application and suitability for use of the antiserum is the responsibility of the user. Behringwerke does not guarantee that the anti-haemoglobin F antiserum will react only with haemoglobin F in all test conditions.

Following enquiries received from Professor B Boettcher a series of tests on antiserum to haemoglobin F was conducted, mainly in the period of March and April, 1983. The repeated results of the tests demonstrates the following conclusions:

Firstly, the antiserum has been adjusted to be specific when reacted against plasma proteins in the agar gel double diffusion (Ouchterlony) technique and immunoelectrophoresis.

Secondly, the antiserum might react with other proteins, e.g. with cell proteins that have entered the plasma.

Thirdly, non-specific immune reactions can be observed under certain conditions due to denaturation of haemoglobin A in adult blood or due to alteration of the relative concentrations of antigen and antibody.

Fourthly, the antiserum against haemoglobin F of Behringwerke, therefore, is not suitable on its own for the identification of foetal/infant blood and adult blood.

This statement is made by Behringwerke AG by duly authorised officers namely Dr. rer. nat. Klaus Störiko, General Manager for Production of Diagnostics and Dr. rer. nat. Siegfried Baudner, Production Manager and Head of the Plasma Protein Research Laboratory.

BEHRINGWERKE
Aktiengesellschaft

Dr Störiko   Dr Baudner

APPENDIX C

COPIED FROM LETTER

OFFICE OF THE DIRECTOR OF PUBLIC PROSECUTIONS
THE NATIONAL PROSECUTING AUTHORITY OF SOUTH AFRICA
IGUNYA LABETSHUTSHISI BOMZANTSI AFRICA
DIE NASIONALE VERVOLGINSGESAG VAN SUID-AFRIKA

CAPE TOWN
Tel: +27 21 487-7000
Fax: +27 21 487 7237

15 Buitengracht Street
Cape Town 8000

P/Bag X900
Cape Town
8000
South Africa

www.npa.gov.za

07 June 2010

Ref No: 9/2/4/5
Enq: Mrs A Lotz

Honourable Minister R Carlisle
Ministry of Transport and Public Works
8th Floor

9 Dorp Street
CAPE TOWN
8001

Dear Honourable Minister

MEETING WITH REPRESENTATIVE OF DRÄGER AND DR KLATZOW

On 4 June 2010 I had a meeting with relevant role players which, inter alia, included representatives of Dräger South Africa, to discuss issues in regard to the Dräger. I was surprised to be informed that you plant to meet later today with a representative of Dräger Germany, South African Breweries and Dr Klatzow.

I am not aware of your meeting's agenda. Despite this, I consider it necessary to draw your attention to recent developments which you may not be aware of. Once you have considered these, you are urged to reconsider your meeting. The relevant developments are as follows:

1. When the recent controversy surrounding the accuracy and legitimacy of the Dräger apparatus arose, it was agreed between my office and the Legal Aid Board to run a so-called 'test case' regarding Dräger in the Wynberg magistrate's court. It is envisaged that the matter will be heard from the 22 June 2010. This case will allow the State to present all relevant evidence and legal arguments in favour of the legitimacy and accuracy of Dräger, while the defence will be able to present the contrary. The court will pronounce on the issue either way. Hopefully, the controversy will then be resolved authoritatively and transparently, in the interests of proper law-enforcement.

2. As part of this process, lawyers that will represent the accused in this matter requested to be provided with a Dräger machine. They wanted it to enable their expert to conduct his own tests, no doubt so that he could present the results as part of the defence case in the subsequent trial. It is no secret that Dr Klatzow has been appointed to conduct these tests. He is thus on brief from the defence. He is for all practical purposes a defence witness in an impending criminal matter. If your proposed

meeting is to include Dräger representatives and experts, this will mean that Dr Klatzow will be engaging in discussions with state witnesses. This presents an unacceptable state of affairs in a criminal case. It places Dr Klatzow in a conflict of interest situation. It might very well jeopardize the up-coming test case. It may even have consequences for the future use of the Dräger in the Western Cape.

3. I was informed at the meeting on 4 June 2010 that Dr Klatzow seeks to gain some business advantage out of the present process: apparently he offers to provide his or his company's services to analyse blood-alcohol samples for the State, on the basis that Dräger is unreliable and there are backlogs with State laboratories. If this is correct, his status as an independent scientist who can provide advice for its own sake would appear to be compromised.

I suggest that the matter can be resolved as follows. If Dr Klatzow has legitimate concerns about the accuracy of the machine, he should provide the prosecuting authority with a report in this regard. This would be in accordance with the practice in a criminal trial that involves experts for the State and defence – reports from the experts are exchanged in advance so that each side has the opportunity to evaluate the experts' evidence.

Based on a proper evaluation of Dr Klatzow's report and on the advice of the Dräger experts for the State, it is my office which must ultimately decide on the way forward. This will include deciding whether Dr Klatzow's concerns, after due and proper consultation and evaluation, can be answered. If they cannot, it may well be decided to abandon the test case and rectify whatever Dräger procedures are found to be wanting, before proceeding again with Dräger. On the other hand, if it appears that Dr Klatzow's criticisms are unfounded, then the test case must proceed, in order for the opposing evidence to be properly adjudicated by the court.

I fail to see the purpose of a debate with you between Dr Klatzow and Dräger as it is not they who will decide on the fate of the Dräger. Their interaction will amount to an extra-curial debate between States witnesses and a defence witness about issues which only a court or my office can resolve. As I have indicated above, a meeting between them at this stage is, from a legal point of view, wholly inappropriate.

If you nevertheless insist that the meeting must proceed, I trust that you will ensure that the meeting will not compromise aspects that will at the opportune time best be addressed in a court of law, or during the formal procedure preceding the hearing.

Kind regards

ADV RJ DE KOCK
DIRECTOR OF PUBLIC PROSECUTIONS: WESTERN CAPE

cc: The Premier of the Western Cape, Ms H Zille
Minister of Community of Safety, Mr L Max

APPENDIX D

The Forensic Scientist CC t/a
INDEPENDENT FORENSIC CONSULTANTS
Reg No: CK 972038723

22 Aberdeen Road
RONDEBOSCH 7700
P O Box 747
RONDEBOSCH 7701
SOUTH AFRICA
Tel: 021-6855747
Pvt Line: 021-6850260
Fax: 021-6857748
Faxmail: 0866844499
Cell: 0828072585
david.klatzow@mweb.co.za

14 June 2010

Our Ref : DJK/nsb/
Your Ref :

MR ROBIN CARLISLE
MINISTRY OF TRANSPORT
AND PUBLIC WORKS
PROVINCIAL GOVERNMENT
WESTERN CAPE

PER EMAIL: ministertpw@pgwc.gov.za

Dear Robin

RE: DRÄGER BREATHALYZER TESTS

I am sorry that I have not been able to respond sooner. However, it is probably better that I have had some time to reflect on the letter from Mr De Kock dated 7 June 2010.

Sole Member : Dr DJ Klatzow

Firstly, it needs to be said that Mr De Kock displays a lamentable lack of understanding about the true role of expert witnesses in our legal system. The true purpose of an expert witness is to guide the court. There is no such thing as a 'prosecution witness' or a 'defence expert'. Scientific evidence is not amenable to being of the prosecution or the defence type. Sometimes the evidence given by an expert witness called by the defence may favour the State case or visa versa.

With regard to my so called 'brief' for Legal-Aid, I have become aware during protracted negotiations with them that they are utterly prostrated by bureaucratic inefficiency. In addition, I am not of any mind to waste my time with people who are incapable of making a decision. Hence, I have withdrawn from acting in this matter. It must also be said that at all times I informed the Legal-Aid Board that I would investigate the instrument and that if I was convinced that it was safe to use (in the legal sense), I would then advise them not to contest the matter. I also informed them that my results and views would be made available to the prosecution.

Mr De Kock is naïve to think that a magistrate will be in any way able to evaluate complex scientific evidence. By far the better way would be to have scientists acting for Dräger interacting with independent experts to evaluate the value of the instrument outside the rather artificial climate of a court case.

Mr De Kock's assertion that there could arise a conflict of interest if I were to meet with the Dräger representatives is rather foolish. I have the same interest as De Kock and for that matter yourself, namely to rid the roads of the scourge of drunken drivers. I have, however, another overriding interest, namely to do so using scientific methods that do not run the risk of convicting innocent persons. In my view, the Dräger may fall into the latter category.

By now you will have been aware of the attempts by Mr Rob Brown of Dräger to stave off my requests for supporting literature. One of his principle excuses was that such literature was only available in German. Apart from the fact that the scientific lingua franca is English, this excuse was shown to be palpably false when Brown delivered to me (considerably overdue)

Sole Member : Dr DJ Klatzow

a wad of papers in English. These are of more than passing interest in as much as they raise most of the concerns which I have already discussed with you.

Regarding paragraph 3 of De Kock's letter, I am once again astounded at the crass stupidity of the assertion that 'Dr Klatzow seeks to gain some business advantage out of the present process.' It is not the first time that such assertions have been made. Some twenty six years ago, the same assertion was made regarding the advice given to the State regarding problems with non sterile tubes. For the record, the State introduced these tubes according to my advice only after a series of humiliating court defeats and incidentally I never made a single cent from the blood tubes nor did I attempt to.

If I could point out to you, the backlogs in all the State laboratories are real and are proving devastating to the legal system. The standards of the scientific work emanating from these laboratories are plummeting with the predictable results which adorn our daily newspapers. The request to investigate the setting up of private laboratories came from the Honourable Premier, not from me and it came long before the case from Legal-Aid. I have also been warning about the Dräger for about two years.

With regard to the insinuations that I have a 'financial advantage' and that my 'status as an independent scientist who can provide professional advice for its own sake' is compromised. I reject this with contempt both for the man who made it and for the sentiment it contains. It is ironic that Mr De Kock exposes his double standards with such clarity. Does Dräger not seek to gain financial advantage from the sale of these instruments at exorbitant prices to the State? On Mr De Kocks' analysis, they will be compromised. De Kock's whole attitude is so puerile as to defy any form of logic. It is the same attitude which pervaded the recent Van der Vyver trial for the murder of Inge Lotz. One thing is clear; De Kock has learned not a single lesson from the appalling and embarrassing mistakes made by the prosecution in that protracted display of State incompetence.

I am not alone in the senior forensic community in harbouring doubts about this instrument as a primary and sole prosecution tool. If the instrument

---

Sole Member : Dr DJ Klatzow

should fail in court, the legal and political fall out for you given your stated good intentions would be a travesty. Dräger's mutual reluctance to co-operate with me has a sinister component. If there was nothing to hide, there could be no downside to Dräger.

If De Kock fails to see the purpose of a meeting between Dräger and myself, that is his problem. However, I am disappointed that you cannot see the advantages.

Kind regards

DR DAVID J KLATZOW

cc: Mr De Kock
Mr Van de Vyver
Premier Helen Zille

Sole Member : Dr DJ Klatzow

APPENDIX E

TYPED FROM A COPY OF THE ORIGINAL AFFIDAVIT

IN THE SUPREME COURT OF SOUTH AFRICA
TRANSVAAL PROVINCIAL DIVISION

Case No:

In the matter between:

BUTANA ALMOND NOFOMELA	APPLICANT
and	
THE MINISTER OF JUSTICE	FIRST RESPONDENT
and	
THE SHERIFF, TRANSVAAL	SECOND RESPONDENT

FOUNDING AFFIDAVIT

I, the undersigned,

BUTANA ALMOND NOFOMELA

do hereby make oath and state:

1.

I am a thirty two year old male presently under sentence of death. My execution is scheduled for tomorrow morning, 20 October 1989, at 07h00.

2.

The contents hereof, unless otherwise indicated by the contents are true to the best of my personal knowledge and belief.

3.

I did not commit the murder for which I stand condemned. I repeat my evidence at the trial which led to my death sentence. I confirm the contentions raised therein by myself and on my behalf by my Counsel.

4.

I wish to hereby reveal facts about my past, which I respectfully contend, might very well have had a bearing on my conviction and/or sentence of death had they been known to the trial court, appeal court and the honourable first respondent.

5.

I was a member of the security branch stationed at headquarters in Pretoria from 1981 until my sentence of death on 21 September 1987.

6.

During the period of my service in the security branch, I served under station commander Brig. Schoon. In 1981, I was appointed a member of the security branch's assassination squad, and I served under Capt. Johannes Dirk Coetzee, who was my commanding officer in the field.

7.

Some time during late 1981 I was briefed by Brig. Schoon and Capt. Coetzee at Pretoria to eliminate a certain Durban attorney, Griffiths Mxenge. I was told by these superiors that Mxenge was to be eliminated for his activities within the African National Congress. They instructed me to travel to Durban in the company of Brian Justice Nqullinga, David Tshikalange and Josephe Mamaselela, colleagues of mine in the assassination squad. I was the leader of this group that was to eliminate Mxenge, and initially I was briefed alone. Thereafter, also in Pretoria, Coetzee briefed the four of us together.

8.

Thereafter, Brian, David, Joseph and I travelled to Durban in one car – a Toyota bakkie with a canopy – where we met Coetzee at CR Swart Police Station. Coetzee had travelled to Durban separately.

9.

Coetzee then gave us a photograph of Mxenge, and he furnished us with details as to Mxenge's whereabouts. Coetzee instructed us specifically not to shoot Mxenge, but to kill him with a knife. Coetzee also mentioned that there were vicious dogs where Mxenge lived, and he gave me a poison which he told me to mix with meat and to throw the meat into the yard for the dogs to eat and hopefully die.

10.

That same day, Brian, David, Joseph and I (Brian always drove as he knows Durban well) went to Mxenge's house in the early evening, after I had bought the meat and mixed it with the poison. On the way to Mxenge's house Brian stopped the bakkie for me to get out and walk the rest of the way to the house. Brian and the others waited in the bakkie at a spot which was on the route that Mxenge normally took on his way home from work. The idea was that if he came past, they would attempt to stop him and kidnap him.

11.

I in the meantime went to the house and threw the meat into the yard as planned. I returned to the others, but Mxenge had not yet appeared.

12.

We then returned to the CR Swart Police Station, where we were barracked for the duration of our stay in Durban.

13.

For the next few days we monitored Mxenge's movements. Then on a particular day, we parked our car in the middle of the road that he normally used to return home from work, very close to his home. It was late afternoon. We saw a white Audi approaching from the distance. We then opened the bonnet of the bakkie, pretending that the bakkie was stuck. Mxenge stopped behind the bakkie. He opened his window, and asked whether he could help us. I approached the car, and I said, 'Yes please.' He then switched off his ignition, and at the same time I produced my firearm, a Makarov pistol. I opened his door, and I ordered him to move over so that Brian could get in behind the wheel. Brian got in behind the wheel. Brian had also produced a firearm, and Joseph got into the back of Mxenge's car. I ordered David to follow us in our bakkie, and then I got into the back of Mxenge's car as well.

14.

We had already decided previously that we would take Mxenge to the Umlazi Stadium, and that we would kill him there. At the stadium, more particularly outside the stadium, we all got out and ordered Mxenge to get out as well. We then started assaulting him with kicks and punches, until he fell to the found. We then all stabbed him several times. He immediately died, and we carried on butchering his body. In accordance with our instructions from Capt. Coetzee, we removed Mxenge's items of value like money and a watch in order to simulate a robbery.

15.

From the stadium, where we left Mxenge's body, we returned to the CR Swart Police Station with the bakkie and Mxenge's car. We parked both vehicles in a courtyard parking area at a courthouse alongside the police station. Then I went into the police station to inform Capt. Coetzee that the mission had been completed. I then accompanied Coetzee to Mxenge's car, and Coetzee removed the number plates thereof and fixed false number plates to the car. He then dismissed the other three, and told me to accompany him to Piet Retief that night. He drove Mxenge's car, and I drove Coetzee's service bakkie.

16.

In the early hours of the following morning we arrived at the Piet Retief Police Station, where we were met by certain men in plainclothes. They were expecting us. Coetzee drove Mxenge's car and I followed in the bakkie to the home of one of these men, and the car was parked in the garage. The garage was closed, and we all, the two men, Coetzee and I, set about stripping the car of the spare wheel, radio/tape, booster, sheepskins and tools which were removed from the car, and placed in Coetzee's car. Then Coetzee and one of the men in Mxenge's car, I in the bakkie and the other man in his own car drove towards the Swaziland border. We parked the cars in a plantation. Coetzee then asked me to siphon petrol from the bakkie. I did so, and poured the petrol into a container that I had been carrying in the bakkie.

17.

Coetzee then drove Mxenge's car a distance of some 500 metres into the plantation. He then beckoned me to come with the petrol to the car. He took the petrol and poured it all over the inside of Mxenge's car, and then the outside. He then poured a line of petrol outside the car leading up to the car, and lit the grass in this line. The flames reached the car, and explosive sounds were made as the car burnt. We waited until the fire burnt itself out. We then all returned to Piet Retief leaving the burnt-out shell of Mxenge's car in the plantation.

18.

Coetzee and I immediately returned from Piet Retief to Durban in Coetzee's bakkie.

19.

Some days thereafter, Brian, David, Joseph and I returned to Pretoria in the service bakkie in which we had travelled to Durban in, and Coetzee returned in his service bakkie. We drove in convoy.

20.

The next day was month end. Usually at month's end we had a week off. Before going off, Coetzee handed Brian, David, Joseph and I R1000.00 each, which he said was from Brig. Schoon for successfully eliminating Mxenge.

21.

Before we went off for one week, in fact, immediately after our return from Durban, all the items that had been removed from Mxenge's car and placed in Coetzee's service bakkie in Piet Retief were given to Sgt. Schutte by Capt. Coetzee in my presence, with the instruction that the radio/tape and booster were to be installed in Brig. Schoon's service vehicle. After my return from one week's leave, Schutte remarked to me informally that the radio/tape had been installed into Brig. Schoon's service vehicle.

22.

After my return from one week's leave, Capt. Coetzee informed me that Mxenge's wife is also active in the ANC's activities, and that he might require me to eliminate her as well at some future date. This was the last I heard of this.

23.

I was involved in approximately eight other assassinations during my stint in the assassination squad, and also numerous kidnappings. At this stage, I do not recall the names of any of the victims. Some of the assassinations, four in fact, took place in Swaziland, one in Botswana, one in Maseru and one in Krugersdorp, where the victim was the brother of an ANC terrorist. This terrorist had allegedly shot and killed a policeman in De Weldt. The brother had been working in the United Building Society as a security guard (Krugersdorp branch).

24.

All these missions were performed under different officers in the security branch. Another Capt. Coetzee, Maj. De Kock, Lt. Vermeulen, Col. Cronje are these officers. Their superior was at all times Brig. Schoon, who was at all times aware of these missions.

25.

I am instructed that due to a shortage of time, I cannot here furnish details of these other missions.

26.

I now wish to explain why I have only revealed all this information at this stage. Maj. De Kock visited me with Capt. Naude after my sentence of death. De Kock told me that Brig. Schoon had asked him to convey to me that I was not to reveal anything about my activities as a member of the assassination squad, and he further promised that they will help me out out of this problem. This visit by De Kock and Naude was in 1987. Thereafter other members of the security branch visited me at various intervals. They were Lt. Van Dyk, Lt. Letsatse, Const. Mokalapitsa, Const. Khumalo, and some whose names I don't know. They all brought messages from Maj. De Kock that steps are being taken to get me out of the maximum prison.

27.

Then on 12 October 1989 I received my notice of execution, and on 17 October 1989 a Capt. Khoz** and a certain Lt., both members of the security branch, visited me and informed me that the instruction from Maj. De Kock was that I should take the pain. I then realized that I had been betrayed by my superior officers, who had promised to assist me in getting out of the maximum prison.

28.

It was at that stage that I decided to reveal all of the aforegoing, and I sent a message to the Lawyers for Human Rights to send someone to me to take a statement accordingly and to apply for a stay of my execution.

____

BUTANA ALMOND NOFOMELA

SIGNED AND SWORN TO BEFORE ME AT PRETORIA ON THIS THE 19TH DAY OF OCTOBER 1989, THE DEPONENT HAVING ACKNOWLEDGED THAT HE KNOWS AND UNDERSTANDS THE CONTENTS OF THIS AFFIDAVIT AND THAT HE CONSIDERS THE OATH TAKEN BY HIM TO BE BINDING ON HIS CONSCIENCE.

REFERENCES

BOOKS

Aronson, Elliot. *The Social Animal*. Worth Publishers, 2011

Ashbaugh, David R. *Quantitative-Qualitative Friction Ridge Analysis: An Introduction to Basic and Advanced Ridgeology*. Boca Raton, Florida: CRC Press, 1999

Baker, Mark. *Cops: Their Lives in Their Own Words*. New York: Simon & Schuster, 1985

Bell, Suzanne. *Drugs, Poisons, and Chemistry*. New York: Facts on File, Inc., 2009

Biasotti, A., and J. Murdock. 'The Scientific Basis of Firearms and Toolmark Identification', in David Faigman, David H. Kaye, Michael J. Saks, and Joseph Sanders (eds), *Modern Scientific Evidence: The Law and Science of Expert Testimony*, Vol. 2. West Pub Co., 2002

Broad, William, and Nicholas Wade. *Betrayers of the Truth: Fraud and Deceit in the Halls of Science*. London: Century Publishing, 1983

Brown, Douglas G., and E.V. Tullett. *Bernard Spilsbury: His Life and Cases*. London: George G. Harrap and Co., 1951

Burrard, Sir Gerald. *The Identification of Firearms and Forensic Ballistics*. New Haven: A.S. Barnes and Company, 1962

Chomsky, Noam. *The Common Good*. Odonian Press, 1998

Cooper, W.E., T.G. Schwär, and L.S. Smith. *Alcohol, Drugs and Road Traffic*. Cape Town: Juta, 1979

DeHaan, John. *Kirk's Fire Investigation*, 4th Edition. New York: Prentice Hall, 1997

Denning, Lord Alfred Thompson. *What Next in the Law?* London: Butterworths, 1982

Dershowitz, Alan M. *Reasonable Doubts: The O.J. Simpson Case and the Criminal Justice System*. New York: Simon & Schuster, 1996

Edmond, Gary. 'Pathological Science? Demonstrable Reliability and Expert Forensic Pathology Evidence', in K. Roach (ed.), *Pediatric Forensic Pathology and the Justice System*. Toronto: Queen's Printer for Ontario, 2008

Emsley, John. *The Elements of Murder: A History of Poison*. Oxford: Oxford University Press, 2005

Feigl, F., and V. Anger. *Spot Tests in Organic Analysis*, trans. Ralph E. Desper. Amsterdam: Elsevier, 1966

Fisher, Jim. *Forensics Under Fire: Are Bad Science and Dueling Experts Corrupting Criminal Justice?* New Brunswick: Rutgers University Press, 2008

Gentry, Curt. *J. Edgar Hoover: The Man and the Secrets*. New York and London: W.W. Norton and Company, 1991

Gray, John. *Lawyers Latin: A Vade Mecum*. London: Robert Hale, 2006

Grobler, L. *Crossing the Line: When Cops Become Criminals*. Johannesburg: Jacana Education, 2013

Harris, David A. *Failed Evidence: Why Law Enforcement Resists Science*. New York: New York University Press, 2012

Heard, Brian J. *Handbook of Firearms and Ballistics: Examining and Interpreting Forensic Evidence*. New Jersey: John Wiley and Sons, 1997

Henry, E.R. *Classification and Uses of Fingerprints*. London: His Majesty's Stationery Office, Darling and Sons, 1905

Henssge, Claus, Bernard Knight, Thomas Krompecher, Bernard Madea, and Leonard Nokes. *The Estimation of Time Since Death in the Early Post-mortem Period*. Boca Raton, Florida: CRC Press, 1995

Hoffmann, L.H., and D.T. Zeffertt. *The South African Law of Evidence*, 4th Edition. London: Butterworths, 1994

Houck, Max M., and Jay A. Siegel. *Fundamentals of Forensic Science*. Oxford: Academic Press, 2006

Howard, M.N., Peter Crane, and David A. Hochberg. *Phipson on Evidence*, 14th Edition. London: Sweet & Maxwell, 1990

Imwinkelried, Edward J. *The Methods of Attacking Scientific Evidence*, 4th Edition. London: LexisNexis, 2004

Inbau, Fred E., John E. Reid, Joseph P. Buckley, and Brian C. Jayne. *Criminal Interrogations and Confessions*, 3rd Edition. Baltimore: Williams and Wilkins, 1962

Karch, Steven B. *Pathology of Drug Abuse*. Boca Raton, Florida: CRC Press, 1993

——— (ed.). *Forensic Issues in Alcohol Testing*. Boca Raton, Florida: CRC Press, 2008

Kaye, D.H. 'The Current State of Bullet Lead Evidence', in David L. Faigman, *Modern Scientific Evidence: The Law and Science of Expert Testimony*. Minnesota: Thomson/West, 2007

Kelly, John F., and Phillip K. Wearne. *Tainting Evidence: Inside the Scandals at the FBI Crime Lab*. New York: The Free Press, 1998

Kessler, Ronald. *The Secrets of the FBI*. New York: Broadway Paperbacks, 2011

Klatzow, David, and Sylvia Walker. *Steeped in Blood: The Life and Times of a Forensic Scientist*. Cape Town: Zebra Press, 2010

Knight, Bernard. *Forensic Pathology*, 2nd Edition. London: Edward Arnold, 1991

Loftus, Elizabeth. *Eyewitness Testimony*. Harvard, Connecticut: Harvard University Press, 1996

Macintyre, Ben. *Operation Mincemeat: The True Spy Story that Changed the Course of World War II*. London: Bloomsbury, 2010

McBarnet, Doreen J. *Conviction: Law, the State and the Construction of Justice*. London: Macmillan, 1983

Morton, James. *Bent Coppers: A Survey of Police Corruption*. London: Little, Brown and Company, 1993

Münsterberg, Hugo. *On the Witness Stand: Essays on Psychology and Crime*. New York: Clark Boardman Co., 1908

Pates, Richard, and Diane Riley (eds). *Harm Reduction in Substance Use and High-Risk Behaviour: International Policy and Practice*. New Jersey: Blackwell Publishing, 2012

Pauw, Jacques. *In the Heart of a Whore: The Story of Apartheid's Death Squads*. Southern Book Publishers, 1991

Perrow, Charles. *Normal Accidents Living with High-Risk Technologies*. Princeton, N.J.: Princeton University Press, 1984

Robbins, Louise M. *Footprints: Collection, Analysis and Interpretation*. Illinois: Charles C. Thomas, 1985

Robertson, Bernard, and G.A. Vignaux. *Interpreting Evidence: Evaluating Forensic Science in the Courtroom*. New Jersey: John Wiley and Sons, 1995

Rowe, Walter F. 'Firearms Identification', in Richard Saferstein (ed.), *Forensic Science Handbook*, Vol. II. New York: Prentice Hall, 1988

———. 'Statistics in Forensic Ballistics', in C.G.G. Aitken and D.A. Stoney, *The Use of Statistics in Forensic Science*, Ellis Horwood Series in Forensic Science. Boca Raton, Florida: CRC Press, 1991

Saferstein, Richard (ed.). *Forensic Science Handbook*, Vol. II. New York: Prentice Hall, 1988

Shapiro, H.A. *Medico-Legal Mythology and Other Forensic Contributions*. Johannesburg: Hugh Keartland, 1975

Siegel, J.A., P.J. Saukko, and G.C. Knupfer (eds). *Encyclopedia of Forensic Sciences*. London: Academic Press, 2000

Simpson, Keith. *Forensic Medicine*, 5th Edition. London: Edward Arnold, 1946

Smith, Sir Sydney. *Forensic Medicine: A Textbook for Students and Practitioners*. London: J.A. Churchill, 1955

Smyth, Frank. *Cause of Death: The Story of Forensic Science*. London: Pan Macmillan, 1980

Söderman, Harry. *Policeman's Lot: A Criminologist's Gallery of Friends and Felons*. US: Funk & Wagnalls, 1956

Spitz, Werner U., and Russell S. Fisher. *Medicolegal Investigation of Death*, 3rd Edition. Illinois: Charles C. Thomas, 1993

Summers, Anthony. *Official & Confidential: The Secret Life of J. Edgar Hoover*. London: Corgi Books, 1993

Swanson, Charles, Neil Chamelin, Leonard Territo, and Robert W. Taylor. *Criminal Investigation, 10th Edition*. New York: McGraw-Hill Companies, 2009

Thomas, Gordon. *Secrets and Lies: A History of CIA Mind Control and Germ Warfare*. UK: J.R. Books, 2008

Vaughan, Diane. *The Challenger Launch Decision: Risky Technology, Culture and Deviance at NASA*. Chicago: University of Chicago Press, 1996

Webster, Richard. *Why Freud Was Wrong: Sin, Science and Psychoanalysis*. London: HarperCollins, 1995

Woffinden, Bob. *Miscarriages of Justice*. Philadelphia: Coronet Books, 1989

JOURNAL ARTICLES

Abdool, R., F.T. Sulliman, and M.I. Dhannoo. 'The Injecting Drug Use and HIV/AIDS Nexus in the Republic of Mauritius', *African Journal of Drug and Alcohol Studies* 5 (2), 2006, pp. 107–16

Aronson, Elliot. 'Back to the Future: Retrospective Review of Leon Festinger's "A Theory of Cognitive Dissonance"', *The American Journal of Psychology* 110 (1), 1997, pp. 127–37

Asher, Richard. 'Munchausen's Syndrome', *The Lancet*, 10 February 1951, p. 339

———. 'Munchausen Syndrome', *British Medical Journal*, 19 November 1955, p. 1271

Barnett, P.D., and R.R. Ogle. 'Probabilities and Human Hair Comparison', *Journal of Forensic Sciences* 27 (2), 2 April 1982, pp. 272–78

Biasotti, A. 'Statistical Study of the Individual Characteristics of Fired Bullets', *Journal of Forensic Sciences* 4 (1), 1959, pp. 34–50

Booker, J.L. 'Examination of the Badly Damaged Bullet', *Journal of the Forensic Science Society* 20 (3), 1980, p. 153–62

Bradford, L.W. 'Barriers to Quality Achievement in Crime Laboratory Operations', *Journal of Forensic Sciences* 25 (4), 1980, pp. 902–7

Burd, D.Q., and P.L. Kirk. 'Tool Marks: Factors Involved in their Comparison and Use as Evidence', *Journal of Criminal Law and Criminology* 32 (6), 1942, p. 679

Burke, A.S. 'Prosecutorial Passion, Cognitive Bias, and Plea Bargaining', *Marquette Law Review* 91, Symposium 2007; Hofstra University Legal Studies Research Paper No. 07-31

Cole, Simon A. 'More than Zero: Accounting for Error in Latent Fingerprint Identification', *Journal of Criminal Law and Criminology* 95 (3), 2005, pp. 985–1078

———, William A. Tobin, Lindsay N. Boggess, and Hal S. Stern. 'A Retail Sampling Approach to Assess Impact of Geographic Concentration on Probative Value of Comparative Bullet Lead Analysis', *Law, Probability and Risk* 4 (4), December 2005, pp. 199–216

Connors, E., T. Lundregan, N. Miller, and T. McEwen. 'Convicted by Juries, Exonerated by Science: Case Studies in the Use of DNA Evidence to Establish Innocence after Trial', US Department of Justice, National Institute of Justice, 1996, pp. 51–2

Corry, Janet. 'Possible Sources of Ethanol Ante and Post-mortem: Its Relationship to the Biochemistry and Microbiology of Decomposition', *Journal of Applied Microbiology* 44, 1978, pp. 1–56

Coyle, Ian R., David Field, and Graham A. Starmer. 'An Inconvenient Truth: Legal Implications of Errors in Breath Alcohol Analysis Arising from Statistical Uncertainty', *Australian Journal of Forensic Sciences* 42 (2), 2010, pp. 101–14

Dror, Itiel E., and David Charlton. 'Why Experts Make Errors', *Journal of Forensic Identification* 56 (4) 2006, pp. 600–15

———, D. Charlton, and A.E. Péron. 'Contextual Information Renders Experts Vulnerable to Making Erroneous Identification', *Forensic Science International* 156, 2006, p. 74–8

———, and Robert Rosenthal. 'Meta-Analytically Quantifying the

Reliability and Biasability of Forensic Experts', *Journal of Forensic Sciences* 53 (4), July 2008, p. 900

Faigman, David L. 'Anecdotal Forensics, Phrenology and other Abject Lessons from the History of Science', *Hastings Law Journal* 59, 2008, pp. 979–980

Fisher, Stanley Z. 'In Search of the Virtuous Prosecutor: A Conceptual Framework', *American Journal of Criminal Law* 197, 1988, pp. 204–13

Foley, M.O. 'Police Perjury: A Factorial Survey', PhD Thesis, Graduate Faculty of Criminal Justice, The City University of New York, 2000, NIJ Grant No 98–IJ–CX–0032

Foran, David, Beth E. Wills, Brianne M. Kiley, Carrie B. Jackson, and John H. Trestrail. 'The Conviction of Dr Crippen: New Forensic Findings in a Century-Old Murder', *Journal of Forensic Science* 56 (1), January 2011, pp. 233–40

Freckelton, Ian. 'Science and the Legal Culture', *Expert Evidence* 2 (2) 1993, p. 107

Garrett, Brandon L., and Peter J. Neufeld. 'Invalid Forensic Science Testimony and Wrongful Convictions', *Virginia Law Review* 95 (1), 2009

Gaudette, B.D. 'A Supplementary Discussion of Probabilities and Human Hair Comparisons', *Journal of Forensic Sciences* 27 April 1982, pp. 279–89

Gershman, Bennett L. 'The New Prosecutors', *University of Pittsburgh Law Review* 53, 1992, pp. 393–458

Giannelli, Paul C. 'Independent Crime Laboratories: The Problem of Motivational and Cognitive Bias', *Utah Law Review* 2, 2010, p. 247–66

———. 'The Abuse of Scientific Evidence in Criminal Cases: The Need for Independent Crime Laboratories', *Virginia Journal of Social Policy and the Law* 4, 1997, p. 439

———, and Kevin C. McMunigal. 'Prosecutors, Ethics, and Expert Witnesses', *Fordham Law Review* 76, 2007, pp. 1493–537

Grieve, David L. 'The Identification Process: Attitude and Approach', *Journal of Forensic Identification* 35 (5), 1988, pp. 211–23

Hand, Learned. 'Historical and Practical Considerations Regarding Expert Testimony', *Harvard Law Review* 15, 1901, pp. 40–58

Imwinkelried, Edward T., and William A. Tobin. 'Comparative Bullet Lead Analysis (CBLA) Evidence: Valid Inference or Ipse Dixit?' *Oklahoma City University Law Review* 28, 2003, pp. 43–72

Jonakait, Randolph N. 'Forensic Science: The Need for Regulation', *Harvard Journal of Law and Technology* 4, 1991, pp. 109–91

Joy, Peter A. 'The Relationship between Prosecutorial Misconduct and Wrongful Convictions: Shaping Remedies for a Broken System', *Wisconsin Law Review*, 2006, p. 399

Kirk, Paul L. 'The Ontogeny of Criminalistics', *Journal of Criminal Law and Criminology* 54 (2), 1963, pp. 235–58

Klatzow, D.J. 'Bullet Markings: The Need for Standardised Criteria', Unpublished Paper, 1995

Knight, Sir Bernard. 'Forensic Chemistry in the Dock: Review of *Forever Lost, Forever Gone* by Paddy Joe Hill', *New Scientist*, 15 July 1995, p. 42

Lentini, John J. 'The Evolution of Fire Investigation and Its Impact on Arson Cases', *Criminal Justice* 27 (1), Spring 2012

Leo, Richard A., and Steven A. Drizin. 'The Problems of False Confessions in the Post-DNA World', *North Carolina Law Review* 82, 2004, p. 891

May, Thomas. 'Fire-Pattern Analysis, Junk Science and *Ipse Dixit*: Emerging Forensic 3D Imaging to the Rescue', *Richmond Journal of Law and Technology* XVI (4), 2010, available at http://jolt.richmond.edu/v16i4/article13.pdf

McClurg, Andrew J. 'Good Cop, Bad Cop: Using Cognitive Dissonance Theory to Reduce Police Lying', *University of California-Davis Law Journal* 32, 1999. University of Memphis Legal Studies Research Paper No. 30, available at http://ssrn.com/abstract=1630533

Meadow, Roy. 'Münchausen Syndrome by Proxy: The Hinterland of Child Abuse', *The Lancet*, 13 August 1977

Meares, Tracey L. 'Rewards for Good Behavior: Influencing Prosecutorial

Discretion and Conduct with Financial Incentives', *Fordham Law Review* 64, 1995, pp. 851–921

Mnookin, Jennifer L. 'Of Black Boxes, Instruments and Experts: Testing the Validity of Forensic Science', *Episteme: A Journal of Epistemology* 5 (3), 2008, pp. 343–58

———. 'The Courts, the NAS and the Future of Forensic Science', *Brooklyn Law Review* 75 (4), 2010

———, Simon A. Cole, Itiel E. Dror, Barry A.J. Fisher, Max M. Houck, Keith Inman, David H. Kaye, Jonathan J. Koehler, Glenn Langenburg, D. Michael Risinger, Norah Rudin, Jay Siegel, and David A. Stoney. 'The Need for a Research Culture in the Forensic Sciences', *UCLA Law Review* 58, 2011, pp. 725–79

Neufeld, Peter J. 'The (Near) Irrelevance of *Daubert* to Criminal Justice and Some Suggestions for Reform', *American Journal of Public Health*, 95 (S1), 2005, pp. S107–13

Peterson, Joseph L., John P. Ryan, Pauline J. Houlden, and Steven Mihajlovic. 'Forensic Science and the Courts: The Uses and Effects of Forensic Science in the Adjudication of Felony Cases', *Journal of Forensic Science* 32, 1987, pp. 1730–53

Plüddemann, A., C.D. Parry, A.J. Flischer, and E. Jordaan. 'Heroin Users in Cape Town, South Africa: Injecting Practices, HIV-Related Risk Behaviours, and Other Health Consequences', *Journal of Psychoactive Drugs* 40 (3), 2008, pp. 273–9

Saks, Michael J. 'Banishing the *Ipse Dixit*: The Impact of Kumho Tire on Forensic Science Identification Science', *Washington and Lee Law Review* 57 (3), 2000, pp. 879–900

———, and Jonathan J. Koehler. 'The Individualization Fallacy in Forensic Science', *Vanderbilt Law Review* 61 (1), 2008, pp. 199–219.

Sanders, J. '"Utterly Ineffective": Do Courts Have a Role in Improving the Quality of Forensic Expert Testimony?', *Fordham Urban Law Journal* 38 (2), 2010

Schwartz, Adina. 'Challenging Firearms and Toolmark Identification, Part One', *The Champion*, October 2008, pp. 10–19, www.NACDL.org

Sheppard, Steve. 'The Metamorphosis of Reasonable Doubt: How Changes in the Burden of Proof Have Weakened the Presumption of Innocence', *Notre Dame Law Review* 78, May 2003, pp. 1165–81

Sunstein, Cass. 'The Law of Group Polarization', *Journal of Political Philosophy*, 10 (2), pp. 175–95, 2002

Thompson, William C. 'Beyond Bad Apples: Analysing the Role of Forensic Science in Wrongful Convictions', *South Western University Law Review* 37, 2008, pp. 1027–50

———, and E. Schumann. 'Interpretation of Statistical Evidence in Criminal Trials: The Prosecutor's Fallacy and the Defense Attorney's Fallacy', *Law and Human Behavior* 11 (3), 1987, pp. 167–87

Thomson, M.A. 'Bias and Quality Control in Forensic Science: A Cause for Concern', *Journal of Forensic Sciences* 19 (3), pp. 504–17

Tobin, William A. 'Comparative Bullet Lead Analysis: A Case Study in Flawed Forensics', *The Champion*, July 2008, p. 12–18, www.NACDL.org

Turvey, Brent E. 'Forensic Fraud: Evaluating Law Enforcement and Forensic Science Cultures in the Context of Examiner Misconduct', PhD Thesis, Band University Queensland, Australia, Sept 2012

———. 'Forensic Frauds: A Study of 42 Cases', *Journal of Behavioural Profiling* 41 (1), April 2003

Uchiyama, Tsuneo. 'Toolmark Reproducibility on Fired Bullets and Expended Cartridge Cases', *Journal of the Association for Firearm and Toolmark Examiners* 40 (1), Winter 2008, pp. 3– 46

Vanezis, P. 'Interpreting Bruises at Necropsy', *Journal of Clinical Pathology* 54, 2001, pp. 348–55

Wells, Gary L., and Eric P. Seelau. 'Police Lineups as Experiments: Social Methodology as a Framework for Properly Constructed Lineups', *Personality and Social Psychology Bulletin* 16, 1990, pp. 106–17

NEWSPAPER ARTICLES

Cohen, David. 'He Doesn't Like Women, says ex-wife', *Evening Standard*, 23 January 2004

Cohen, Nick. 'Schooled in Scandal, Judge-Led Enquiries have a Long Track Record of Failing to Criticise Government of Their Day', *Observer*, 1 February 2004

Editorial. 'The FBI Messes Up', *New York Times*, 26 May 2004.

Hsu, Spencer S., Jennifer Jenkins, and Ted Mellnik. 'DOJ review of flawed FBI forensic process lacked transparency', *Washington Post*, 17 April 2012

Jaraline, Cassandra. 'Has Sally Clark's case changed attitudes towards infant death?', *The Telegraph*, 16 March 2008

Sedley, Stephen. 'A Benchmark of British Justice', *Guardian*, 6 March 1999

LEGAL JUDGMENTS, COMMISSIONS AND REPORTS

Bronwich, Michael R. 'Final Report of the Independent Investigator for the Houston Police Department Crime Laboratory and Property Room', 13 June 2007, available at http://www.phdlabinvestigation.org

Centre for Public Policy Evaluation. 'The Family Court: A View from the Outside', Issues Paper No. 3, College of Business, Massey University, October 1998

City of New York Commission to Investigate Allegations of Police Corruption and the Anti-Corruption Procedures of the Police Department: Commission Report, chaired by Milton Mollen, 7 July 1994

Collaborative Testing Services. 'Forensic Laboratory Proficiency Testing Program: Latent Prints Examination', Report No. 9508, Sterling, Virginia

Cork, Daniel L., John E. Rolph, Eugene S. Meieran, and Carol V. Petrie (eds). *Ballistic Imaging*. Washington, D.C.: The National Academies Press, 2008

McGregor, John, John Dingwall, and Gary Dempster. 'Report on Fingerprint Comparison of Mark QI2: David Asbury Case', p. 9, available at www.thefingerprintinquiryscotland.org.uk

Mullin, Chris. 'Evidence to Sir John May's Inquiry into the Guildford and Woolwich Bombings – 1989', available at www.chrismullinexmp.com/speeches/guildford-and-woolwich-bombings

National Research Council of the National Academies. *Forensic Analysis: Weighing Bullet Lead Evidence*. Washington, D.C.: The National Academies Press, 2004

———. *Strengthening Forensic Science in the United States: A Path Forward*. Washington, D.C.: The National Academies Press, 2009

Office of the Inspector General. 'A Review of the FBI's Handling of the Brandon Mayfield Case: Special Report', March 2006, available at http://www.justicegov/org/special/s060/final.pdf

Opinion of Lord Wheatley, *Shirley Jane McKie* v. *Strathclyde Joint Police Board*, 24 December 2003, paragraph 15

Report of the Commissioner the Hon. Mr Justice T.R. Morling/Royal Commission of Inquiry into Chamberlain Convictions, Canberra: Government Printer, 1987

Scottish Parliament Justice 1 Committee Report: Inquiry into the Scottish Criminal Record Office and Scottish Fingerprint Service, 2006, available at http://archive.scottish.parliament.uk/business/committees/justice1/reports-07/j1r07-03-vol1-01.htm

UNAIDS. 'Aids Epidemic Update', Geneva, 2009

CASE AND STATUTORY LAW

Commonwealth v. *Loomis*, 113 A. 428, 430 (PA. 1921) and 110 A. 257 (PA. 1920)

Davie v. *Edinburgh Magistrates* [1953] SC 34 at pp. 545–6

Daubert v. *Merrell Dow Pharmaceuticals Inc*. (509 US 579 (1993))

Larg Heinke v. *Andrew Hubbard*, Case No. 966/2003, High Court of South Africa, Cape of Good Hope Provincial Division, 26 January 2007, Foxcroft, J

Michigan Millers Mutual Insurance Corporation v. *Benfield* 140, F. 3d 915 (11th Circular 1998, available at http://tinyurl.com/82c3d86)

Maritime and General Insurance Company Ltd v. *Skye Unit Engineering (Pty) Ltd* (1989 (1) SA 867 at 887 A.)

Maryland v. *Rose*, No K06-0545 (Balt County Gr Court 2007 (available at http://www.latent-prints.com/images/state%20Rose%20K06-0545.pdf)

Road Traffic Ordinance: Section 140 (4), inserted by Section 14, Ord 7 of 1968 T; Section 13, Ord 27 of 1968 C; Section 6, Ord 18 of 1972 (N) and Section 14, Ord 8 of 1968 (O)

R v. *McIlkenny and Others*, Birmingham Six Appeal Court Judgment, HOJO Scotland

R v. *Ward*, 1993, 96 Crim App 1, 68, UK

State v. *Chun* (available at http://www.judiciary.state.nj.us/mcs/case_law/state_v_chun.pdf)

State v. *Hazel Joyce Harrison*, 6/2/85 Case SH2/39/85, p. 66.

State v. *Hendricks* [2011] 4 All SA 443 (WCHC)

State v. *Malindi*, 1983, (4) SA 99

State v. *Nala*, 1965 (4) SA 360 (A)

State of Maryland v. *Bryan Rose*, In the Circuit Court for Baltimore County, Case No: K06–545

United States v. *Harvard*, 117 F. Supp. 2d 848 (SD. Ind. 2000)

United States v. *Llera Plaza II*, 188 F. Supp. 2d 549 (ED Penn 2002)

United States v. *Rose*, No CCB-08-0149, 2009 WL 4691612 at 3 (D.MD. Dec. 8, 2009)

WEBSITES

http://paulviking.websitetoolbox.com/post/FBI-ordered-McKie-case-swept-under-carpet 969666

http://www.fbi.gov/pressrel/pressrel04/mayfield052404.htm

http://www.fbi.gov.pressrel/pressrelos/bullet_lead_analysis.htm

http://www.innocenceproject.org/

http://www.propublica.org/article/reversal-of-fortune-a-prosecutor-on-trial

http://www.psmag.com/legal-affairs/why-fingerprints-arent-proof-47079/

http://www.TruthInJustice.org

INDEX

Abels, Anthony 40–45
ACE-V methodology 77, 185–187
addiction 146–149, 151–154
AFIS *see* Automated Fingerprint Identification System
African National Congress *see* ANC
AFTE *see* Association of Firearm and Tool Mark Examiners
AFTE Journal 166
AIDS 151–152
AK-47 assault rifle 111
Akili, Talibah 172
alcohol 22–25, 98, 131–136, 226 *see also* Dräger breathalyser
Alcotest 7110 Mark III *see* Dräger breathalyser
Altbeker, Antony 78
Alverston, Lord Richard 11
America *see* United States
American Civil War 146
amphetamines 146, 151
ANC 83–84, 205
Anderson, Sandra 172
Anstey, Len 97
Anthony, Donna 94
antibodies 102–103
apartheid-era atrocities *see* atrocities in apartheid-era
arsenic 7–8, 20
Asbury, David 53–56, 68
Ashbaugh, David R. 47–48
Asher, Richard 89
Association of Firearm and Tool Mark Examiners (AFTE) 161, 164, 166, 223
atrocities in apartheid era 36–37, 39–45, 95–98, 177–178, 184, 195–196
Atwell, Kevin 24
Australia 100–106, 142–143
Australian Journal of Forensic Sciences 142
Automated Fingerprint Identification System (AFIS) 72, 75
Aviv, Juval 68
Ayers Rock, murder at 100–106

Bacasnot, Concepcion 172
bacterial meningitis 93–94
ballistics 5–6, 40–45, 110–114, 154–167, 200–201, 224, 232
Bandner, Dr 103
Banks, Isabella 7–9
Barnett, P.D. 109
Barritt, Denis 100
Barry, John 58
Bartholomew, Bruce 34–36, 198, 209, 214–215
Battle, Jeffrey 72
Baxter, Simon 102, 104–106

Bayer 146, 147
Bayes, Thomas 84–85
Beeches, Thomas 89
Behn, Michael 112–113
Bekker, Daan 81
Bent Coppers 183
Benzedrine 148
Benzien, Jeffrey 40–45
Berman, Peter 141
Bernard Spilsbury 9
Berry, John 56
Berson, David 22
Betrayers of the Truth 219
'Beyond Bad Apples' 206–207
bias 26, 39, 48–49, 73–76, 108, 222
Biasotti, Al 162
Biko, Steve 37, 39, 176
Birkenhead, Lord 10
Birkett, Norman 188–189, 191
'Birmingham Six' 116–120, 182–183
Bisbing, Richard E. 109
bite marks 175, 201
Blackstone, William 115
Blair, Tony 99–100
blood-alcohol testing *see* alcohol
blood spatter 38–39, 101–105, 173
bloodstain analysis 101–104, 224
bodily fluids *see* serology
Bodziak, William 36, 175, 198, 215
Boettcher, Barry 106
bombings in UK, 1974
    see 'Birmingham Six'
Bonhoeffer, Dietrich 121
Bonner, Raymond 198
Booker, J.L. 159–160
Book of the Dead 1
Boshoff, Wimpie 78
Bosnian proverb 95
Boyle, Alan 56
Bozalek, Lee 96
Bradford, L.W. 224
Brett, David 101
Britain 100, 146, 227
Broad, William 219
Brontë, Robert Matthew 13–15
Brooklyn Law Review 73
Brown, John 172
Brown, Rob 138
Browne, Douglas G. 9
bruising 14–18
Brunner, Hans 105–106
bullets 5–6, 110–114, 157–167, 200–201,
    224
Burd, David Q. 159
Burns, Robert 53

cadmium 136
calibration 136, 140
Callaghan, Hugh 116–117
Callaghan, Tom 172
Cameron, Elsie 12–18, 20–21
Cameron, James 105–106
Camus, Albert 115
canine-detection methods 128, 172
Cannery Row 2, 148
Cannings, Angela 94
Cape Argus 138
Cape Times 6
Carlisle, Robin 136–138, 140, 144
Cawley, Bernard 216
Central Intelligence Agency *see* CIA
CESDI *see* 'Confidential Enquiry into
    Stillbirths and Deaths in Infancy'
Chamberlain, Alice Lynne 'Lindy'
    100–106
Chamberlain, Azaria 100–102, 105
Chamberlain, Michael 100–101
Champion, The 166
Charlton, David 74
chemicals, imported 136
China 136, 145–146
Chomsky, Noam 60
Churchill, Winston 10, 71
CIA 148–149
civil trials 32, 211
Clark, Christopher 91, 93–94
Clark, Harry 91
Clark, Sally 87, 91–94
Classification and Uses of Fingerprints 47
Clift, Alan 27–28, 171, 175, 199

Coca-Cola 148
cocaine 147–148, 151, 153
Cocks, Sergeant 105
Coetzee, Dirk 177–178, 184
cognitive bias *see* bias
cognitive dissonance 214–216
Cohen, David 88
Coldicott, Elizabeth 12
Cole, Simon A. 57
collaboration between scientists 166–167, 219, 221–222
Collins, Harry M. 145
comparison microscopy 111, 157–167, 224–225
conduction 122
confessions 119–120, 201
'Confidential Enquiry into Stillbirths and Deaths in Infancy' (CESDI) 91–92
Connelly, Michael 195
Constitution of South Africa 26–27
convection 122
Cooper, Lord President 67, 192
corporate culture 60–61, 69–70, 170, 207–209
Corry, Janet 24
costs 2, 32, 215
cot death *see* Sudden Infant Death Syndrome
Council for Scientific and Industrial Research (CSIR) 124–125
courts 1–2
Cowans, Stephan 82
Coyle, Ian R. 142
crazing of glass 125
Criminal Interrogation and Confessions 201
Criminal Investigation 157
criminal justice system 207–209
Crippen, Cora 9–11
Crippen, Hawley Harvey 9–11
Crislip, Mark 7
cross-examination 174, 188–189
Crossing the Line 149
Crucible, The 93
CSI effect 107–114, 179, 200, 229
CSIR (Council for Scientific and Industrial Research) 124–125
Culshaw, Edward 155

DA *see* Democratic Alliance
Daniels, Judge 83
Daoud, Ouhnane 72, 76
Daubert trilogy 187–188
Daubert v. *Merrell Dow Pharmaceuticals Inc.* 129, 184–185, 187–188, 190, 211
Davie v. *Edinburgh Magistrates* 192
DeHaan, John 123
De Kock, R.J. 138
Democratic Alliance 136–137
denial, mentality of 60–61, 131
Denning, Lord Alfred Thompson 119, 181–184, 194
Dershowitz, Alan 215
detention without charge 100, 117
Dew, Chief Inspector 9–10
dingoes 100–101, 105–106
Disraeli, Benjamin 83
Dixon, Timothy 172
DNA-analysis techniques 11, 82, 110, 173–175, 193, 198, 200
dogs *see* canine-detection methods
Dotson, Gary 174
Dowd, Michael 215
Downer, Billy 138–139, 141
Doyle, Sir Arthur Conan 18
Dräger breathalyser 137–144
Drayton, Janet 119
driving while intoxicated (DWI) *see* alcohol
Dror, Itiel E. 73–74, 212
drugs 34, 145–154, 215–216
drunken driving *see* alcohol
Du Preez, Max 178
DWI (driving while intoxicated) *see* alcohol

Edmond, Gary 210–211
Edwards, Graham 32
Elliot, Walter 119
Encyclopedia of Forensic Sciences 107

England *see* Britain
Evening Standard 88
experience of witnesses *see* 'in my experience'
experts 2–3, 6, 32, 67–68, 168–180
eyewitness identifications 200–201, 212
Eyewitness Testimony 201

Fain, Charles 110
FBI (Federal Bureau of Investigation) 69–70, 72–76, 110, 112–114, 178, 190, 207, 222
fertiliser 136
Festinger, Leon 214
fibre analysis 201
Field, David 142
fingerprints
  ACE-V methodology 77, 185–187
  background to 46–48, 79–82, 164, 187–188, 190–192, 200, 212
  bias 48–49
  fraud 49–52
  Inge Lotz case 77–82
  Mayfield case 71–77
  McKie case 53–70
Finlay, Justice 15
firearms 5–6, 40–45, 110–114, 154–167
fire investigation 121–130, 186, 191
Fischer, Russell S. 17
Fish, Pamela 172, 198
Fisher, Jim 114, 169
Fitzpatrick, John 172
footprints *see* shoe-print comparisons
Footprints 175
Foran, David 11
Forensic Fraud 204–205
Forensic Medicine 16–17
Forensic Pathology 99
Forensic Science Handbook 109, 159
Forensic Science International 74
Forensic Science Society 167, 223
Forensics Under Fire 114, 169
fraud 48–52, 171–180, 204–210, 212–217, 219–222
Freud, Anna 88
Freud, Sigmund 147
Fruit of a Poisoned Tree 78
Fundamentals of Forensic Science 108
Furlong, Mary 172

Galileo Galilei 71
Galt, Hugh Millar 13
Galton, Sir Francis 46–47
Galvin, G. 100
gangs 69–70
Gardner, G.Y. 164
Garrett, Brandon L. 173–174, 179, 192–193
gas chromatography 23, 118, 135, 150
gas chromatography mass spectrometry (GCMS) 118, 128, 150
Gaudette, B.D. 109
Gauntlett, Jeremy 41
Gawie 124–125, 130, 176
GCMS *see* gas chromatography mass spectrometry
Geddes, Alister 54–55
General Electric v. *Joiner* 187
Gentry, Curt 70
Germany 19, 142
Gilchrist, Joyce 192–193, 198
Gill, Dorothy 24
Goldwyn, Samuel 22
Graham, Malcolm 56
Gray, Thomas 88
Griess test 116–119
Grieve, David L. 49, 56, 64–68
Grobler, Liza 149
Grunbaum, Benjamin W. 225
Guantanamo Bay detention camp 100
'Gugulethu Seven' 205–207
guns *see* firearms

hair analysis 109–110, 201
Hakamada, Iwao 6
Halsey, Margaret 131
Hand, Learned 168
handwriting comparison 83–84, 225
Harding, Dr 105–106
Harding, David 172

Harm Reduction in Substance Use and High-Risk Behaviour 151
Harms, Louis 184
Harvard see United States v. *Harvard*
Hattingh, Hannes 83–84
Heard, Brian J. 162–164, 166
Hembury, Kevin 215
Henry, Edward R. 47
Henry Classification System 47
Henry's law 141
Henssge nomogram 31–32
heroin 146–147, 151–152
Herschel, Sir William James 3, 47
Higgins, Wallace 172
Higgs, Douglas 119
Hill, Paddy 116–117
Hitchens, Christopher 168
HIV/AIDS 151–152
Hoffman, G. 43–44
Hoffman, Paul 137–139
Hofmann, Albert 148
Holmes, Warrant Officer 33
Hoover, J. Edgar 69–70
Houlden, Pauline J. 229
Houston Police Department Crime Laboratory 179, 207
Hunter, Gerry 116–117

IAAI *see* International Association of Arson Investigators
IAFIS *see* Integrated Automated Fingerprint Identification System
identifications by eyewitnesses 200–201, 212
immune system 16, 102–103
independence of forensic services 227–228
India 146
infallibility 74–75
information, suppression of 166–167, 235, 237–238
'in my experience' 4, 7, 211
innocence, presumption of 99–106, 115, 193–194, 230
Innocence Project 175, 198, 200, 214, 217
Inquest Act 26
insurance industry 4–5, 121–124, 126–128
Integrated Automated Fingerprint Identification System (IAFIS) 72, 75
International Association of Arson Investigators (IAAI) 129
Interpreting Evidence 84–85
In the Heart of the Whore 178
Ireland 116–117, 119
Irish Republican Army (IRA) 116–117
Isaacs, Arthur 188–189
Italy, invasion of 18–19
Ivanov, Eugene (Yevgeni) 184

J. Edgar Hoover 70
Johnson, Lyndon B. 69
Jonakait, Randolph 224–225
Jones, Anthony 104, 106
Journal of Forensic Identification 49
Journal of Forensic Sciences 11, 109
Journal of the Forensic Science Society 160, 167
Joy, Peter A. 199
judges 1, 20–21, 174, 180–194, 211–212, 226–227
justice, concept of 1–3, 65–67, 115–116
Justitia (Roman goddess) 1

Kaminsky, Sam A. 172
Keeler, Christine 184
Kelvin, Lord 3
Kemp, Dennis 22, 25
Kendall, Henry 10
Kennedy, Robert F. 195
Kentridge, Sydney 176
Kessler, Ronald 69
King, Martin Luther, Jr 181
King, Robert Merton 187
King, William 138–139
Kirk, Paul L. 159, 202–203
Kirk's Fire Investigation 123
Knight, Sir Bernard 90
Knobel, Gideon 43–44
Kobrin, Raymond 33
Kock, Colonel 38–39, 173

Kotze, Nico 79, 81
Kondile, Sizwe 177
Kriel, Ashley 39–45, 98
Krige, Joel 96
Krone, Ray 175
Kuhl, Joy 102–106
Kumho Tire Co. v. *Carmichael* 187

laboratories 113, 135–137, 219, 222, 224–226, 230
Lacassagne, Alexandre 157
Lady Justice 1
Lancaster, Allison 172
Lang, Dr 37
Lansky, Meyer 70
Larg Heinke v. *Andrew Hubbard* 136
laudanum 148
law enforcement *see* police
Legal Aid 138, 223
Legal Resources Centre 96
Le Neve, Ethel 9–10
Lengisi, Bongile 136
Lévi-Strauss, Claude 71
Lewis, Sir George Cornewall 8
Liebenberg, Linda 36
line-ups 200–201, 212
Locard's exchange principle 54
Lockerbie air disaster 68
Loftus, Elizabeth 201
Log from the Sea of Cortez, The 2
Lotz, Inge 34–36, 77–82, 150–151 *see also* Van der Vyver, Fred
LSD 148
Luger pistol *see* Parabellum 9mm pistol

Macbeth 61–62
Macintyre, Ben 19
Mackenzie, Robert 56
MacLeod, John 58–60
Mafia 70
Magna Carta 115–116
Maguire family 119–120
Malone, Michael P. 110, 113
Mandrax 34, 146, 149, 173
Maritime and General Insurance Company Ltd v. *Skye Unit Engineering (Pty) Ltd* 192
Maritz, Frans 36
Maryland v. *Rose* 185
mass spectrometers 150
May, Sir John 119
May, Thomas 130
Mayfield, Brandon 71–77, 185–186, 200, 207, 220
McBride, Fiona 55, 61
McDade, James 116–117
McIlkenny, Richard 116–117
McKenna, Anthony 55
McKie, Iain 54, 60
McKie, Shirley 52, 54–68, 95, 179, 186, 200
McNamee, Danny 60
McPherson, Hugh 54–55
Meadow, Samuel Roy 87–94
'Meadow's Law' 93–94
Medical Research Council (MRC) 151
Medico-Legal Mythology and Other Forensic Contributions 20
Meehan, Brian W. 173
melamine 136
'Mellanby effect' 134
Mengele, Josef 70
Merck 145
methaqualone 146, 151
Mfolozi, Sipho 169
Michael, Glyndwr 19
Michigan Millers Mutual Insurance Corporation v. *Benfield* 129
Mihajlovic, Steven 229
Miscarriages of Justice 117
Mitchell, Derek 138, 143
MK *see* Umkhonto we Sizwe
MK-Ultra 148
Mnookin, Jennifer L. 73–74, 143, 166, 185, 187, 190, 202
Model Rules of Professional Conduct 200
Mollen Commission 215
Montagu, Ewen 18
More, George 55
Morling, T.R. 100, 103–106

morphine 145–147, 152
Morrison, Gene 173
Mortimer, John 50
Morton, James 183
MRC *see* Medical Research Council
Mullin, Chris 183
'Münchausen Syndrome by Proxy' 89–91
Münsterberg, Hugo 201
Murdock, John 162

National Academy of Sciences, US 163–164 *see also Strengthening Forensic Science in the United States*
National Fire Protection Association *see NFPA* 921
National Prosecuting Authority 137, 170, 199, 213–214
NCIS see CSI effect
Ndzima 95–98
Neethling, Lothar Paul 5, 34, 70, 149, 176–178, 218
negative corpus hypothesis 123, 169
Nel, Professor 97–98
Nel, Jurie 25
Neufeld, Peter J. 173–174, 179, 192–193
neutron-activation method 112
New York Police Department (NYPD) 215–216
NFPA 921 123, 124
Nietzsche, Friedrich 30, 78
Nifong, Mike 173
ninhydrin 48
Noah 131
Nofomela, Butana Almond 177–178
Ntela, Mr 199
NYPD *see* New York Police Department

'Ode on a Distant Prospect of Eton College' 88
Official and Confidential 70
Ogle, R.R. 109
On the Witness Stand 201
Operation Mincemeat 19
opium 145–146, 148
organisational theory *see* corporate culture
organised crime 70
Orr, Wendy 37, 45
Otto, Charlene 171

Paarl 50
paint comparisons 224–225
Pan Am flight 68
Parabellum 9mm pistol 165
Parabellum bullets 114
Parke-Davis 147
Parry, Charles 151, 153
Pasteur, Louis 226
Patel, Trupti 94
pathologists *see* state pathologists
Pauw, Jacques 178
Pemberton, John S. 148
perjury 95
Péron, Ailsa E. 74
Perrow, Charles 207
Peters, Charles 112
Peterson, Joseph L. 229
Phipson on Evidence 68, 99
phosphorous poisoning 19
photographs 26–27, 129, 199
plasma serum *see* serology
police 27, 32–33, 37–45, 115–116, 149–153, 170–171, 195–197, 201–207, 212–217, 218–224
Pollack, Louis 190
post-mortem interval 30–32
Power, Billy 116–117
Preece, John 27–28
preservation of human body 20
Prevention of Terrorism Act 117
prisons 100, 137, 152 *see also* detention without charge
Profumo, John 184
prosecutors 15, 26–29, 32–34, 195–200, 208–209, 212–217
prosecutor's fallacy 84–87, 94
Publilius Syrus 46
Purkinje, Jan Evangelista 46

quality control 224–226, 230
Quantitative-Qualitative Friction Ridge Analysis 47–48

radiation 122
Randolph-Quinney, Patrick 85–87
raw data 93, 230
reasonable doubt 99–106, 115, 193–194, 230
regulation of forensic science 223–224
Reid technique 201
Reinsch test 8
Richmond Journal of Law and Technology 130
Ricketts, Ed 2, 148
rifled weapons 156–157
rifles 110–111
Rivera, Michael 110
Road Traffic Ordinance 24
Robbins, Louise 173, 175, 198, 209
Rose, Robert 112
Ross, Marion 53–55
Rouse, Alfred Arthur 188–189
Rowe, Walter 159–160, 164
Rumpole of the Bailey 50
Rutherford, Ernest 3
R v. *A.A. Rouse* 188–189, 191
R v. *Seddon* 20
Ryan, John P. 229
Ryder, Paul 82

Saayman, Gert 36, 169
SAB *see* South African Breweries
Sade, Marquis de 154
Sagan, Carl 154
Sandburg, Piet 126–127
Sanders, Joseph 193
Sax, Michael 211
Schumann, Edward 85
Schutz, Judge 6
Schwär, Theo 43–44
Schwartz, Adina 166
scientific culture 219, 222–226
scientific illiteracy 11, 15, 44, 108, 115, 180, 191 *see also* training
Scott, Andrew 105
Scottish Criminal Record Office (SCRO) 54–68, 186
Second World War 18–19, 148
Secrets and Lies 148
Secrets of the FBI, The 69
Selebi, Jackie 149
serology 27–28, 175
Sertürner, Friedrich 145
Seventh-day Adventist Church 101
Shapiro, Hillel Abbe 19, 20, 24
Sheppard, Steve 194
shoe-print comparisons 34–36, 175, 201, 214–215
shotguns 111
SIDS *see* Sudden Infant Death Syndrome
silver nitrate 48
Simpson, Keith 15–16
Simpson, O.J. 36
Skuse, Frank 116–120, 182
Smethurst, Thomas 7–9
Smith, F.E. (later Lord Birkenhead) 10
Smith, Lionel Shelsey 96–97
Smith, Sir Sydney 16–17
'Snaggle Tooth Killer' 175
Söderman, Harry 168
Sohege, Jürgen 141–142, 144
Sophocles 7, 218
Souder, Susan M. 77
South African Breweries (SAB) 137–138, 140, 142
South African Law of Evidence, The 99
South African Medical Journal 19–20
South African Medical Research Council (MRC) 151
Spain 19, 72–73
spalling 124–125
Spar supermarket, Orkney 126
Spilsbury, Sir Bernard 9–21, 45, 93, 229
Spitz, Werner U. 17
Spot Tests in Organic Analysis 117
Staphylococcus aureus 93–94
Starmer, Graham A. 142
Starrs, James E. 3

Star semi-automatic 0.22-calibre pistol 41
Star Trek IV 30
state pathologists 25–29, 30–39, 45, 119–120, 170–171, 206
State v. *Chun* 143–144
State v. *Hendricks* 138–144
State v. *Malini* 192
State v. *Nala* 191–192
State v. *Van der Vyver* *see* Van der Vyver, Fred
statistics 83–94, 163
Steeped in Blood 78, 127, 173
Steinbeck, John 2
Steward, Larry F. 173
Stewart, Charles 55
Stoney, D.A. 164
Störiko, Dr 103
Strauss, Heinrich 218
Strengthening Forensic Science in the United States 107–109, 165–166, 171, 175, 187, 193, 222
strychnine 153
Study in Scarlet, A 18
substance abuse *see* drugs
Sudden Infant Death Syndrome (SIDS) 90–94
Summers, Anthony 70
suppression of information 166–167, 219, 221–222
Surprising Adventures of Baron Münchhausen, The 89
Sutton, Josiah 208
Swann, Peter 50, 56, 58–59
Swartz, Constable 81–82

Taliban 72
Taylor, Alfred Swaine 7–9
tea 146
technicians 218–219
television, impact of *see CSI* effect
terrorism 72–73, 99–100, 117
Theunissen, Carien 171
thin-layer chromatography 119, 150
Thomas, Gordon 148
Thompson, William C. 85, 179, 206–207
Thomson, M.A. 230
Thomson, William 167
Thorne, Norman 12–15, 18, 20–21
time of death 30–32
Tipple, Stuart 103
Tobin, William 113
Toms, John 155
training 108–109, 127, 166, 187, 209, 222–227 *see also* scientific illiteracy
transposed conditional *see* prosecutor's fallacy
truth 2, 95–98
Truth and Reconciliation Commission 206
Tucker, Dr 37
Tullett, E.V. 9
Turner, Cora (was Kunigunde Mackamotski, later Cora Crippen) 9–11
Turvey, Brent E. 204–205, 218, 220–221
Twain, Mark 83
Twigg, Lieutenant 34
tyre prints 35, 160, 201

Uchiyama, Tsuneo 161–162
UCLA Law Review 166
Uluru, murder at 100–106
Umkhonto we Sizwe 205
United Kingdom *see* Britain
United States 69, 100, 143, 146, 187, 200, 222
United States v. *Harvard* 187–188, 189–190
United States v. *Llera Plaza II* 190
US *see* United States

Van der Vijver, Christinus 138–139, 141
Van der Vyver, Fred 34–36, 77–82, 150–151, 171, 194, 197–198, 208–209, 214–215
Vanezis, P. 17
Van Heerden, T.J. 96
Van Nieuwenhuizen, Solly 25
Van Wezel, Nic 25
Vaughan, Diane 207

Vellema, Jeanine 39
Vermeulen, Koos 177
Viljoen, Neels 25, 136
Virginia Law Review 173
Visser, Willem 42–44
Von Bülow, Claus 215
Von Münchhausen, Karl Friedrich Hieronymus Freiherr 89
Vrye Weekblad 178

Wade, Nicholas 219
Walker, Johnny 116–117
Wallace, J. 61–62
Ward, Stephen 184
Weaver, Tony 205
Wells, Gary 200
Wertheim, Pat 56–57, 59, 62, 67–68, 82
West, Michael 198
What Next in the Law 182–183
Wheatley, Lord 60
White, Ellen G. 204
Whitehurst, Frederic 110, 113, 172
'white pipe' 146
Wilkinson, Andrew 139, 142, 144
Wilks, Sir Samuel 8
Will, Helen 27–28
Williams, Alan 93–94
Willingham, Cameron Todd 127
Willis, Ernest Ray 127
Woffinden, Bob 117
Wolfinger, Raymond 46
Woodall, Glen 209
World War II 18–19, 148
Wright, C.R. Alder 146

Young, Anthony 194

Zain, Fred 173, 198, 209
Zeelenberg, Arie 57–59, 82
Zuma, Jacob 170, 216